Home Gardening

Home Gardening

M.L. Chadha
K.L. Chadha
Roohani Pal
Satish Kumar Sain

Kruger Brentt
Publishers
2025

Kruger Brentt Publishers UK. LTD.
Company Number 9728962

Regd. Office: 68 St Margarets Road, Edgware, Middlesex HA8 9UU

ISBN 978-1-78715-004-1 (Hardbound)

Disclaimer:

For information on all our publications visit our website at http://krugerbrentt.com/

Prof. M. S. Swaminathan
Founder Chairman, M S Swaminathan Research Foundation
Third Cross Street, Taramani Institutional Area
Chennai – 600 113 (India)
Email: swami@mssrf.res.in

MSSRF

Foreword

The current population of India is around 1.15 billion and on its current growth trajectory, the population is expected to reach nearly 1.53 billion by the year 2030. Malnutrition and Anemia in young children and women are substantial problems in India. India alone is home to 40% of the world's malnourished children and 35% of the developing world's low-birth-weight infants. Twenty four percent of the children here are severely stunted and 16 are underweight. One of the easiest way of ensuring access to a healthy diet that contains adequate macro- and micronutrients is "Home Gardening"-an age-old practice to supply a diverse range of fruits and vegetables. Home garden models based on vegetables rich in vitamin-A, protein, iron and iodine, would supply vegetables for the whole family throughout the year to ensure nutrition security. The broad diversity of horticultural crop species allows year-round production, employment and income. Intensive horticulture on small plots, making efficient use of limited water, land resources with a considerable yield potential to provide up to 50 kg of fresh produce per m2 per year depending upon the technology applied. In addition, due to their short cycle home gardens provide a quick to emergency needs for food and provide a quick return to meet a family's daily cash requirements. Leafy vegetables are particularly perishable and post-harvest losses can be reduced significantly when production is located close to consumers. No matters if they are urban or peri-urban, all gardens benefit from preplanning and design. There are also good reasons for home garden interventions to activity promote women empowerment. The climate has become increasingly unpredictable, but a diverse garden is a resilient garden. In the context, vegetable crops which fit well in almost all farming system can play an important role not only the economic welfare of the people but also but also help in fighting hunger, malnutrition and ecological imbalance. To meet the challenge of huge demand of vegetable for reducing hunger, malnutrition, & poverty, the productivity and production has to be enhanced at all levels including farmers' fields, urban, peri-urban and home garden. The technological interventions developed in recent past can boost and diversify vegetable production and consumption.

The publication on Home gardening topic are scattered in literature and comprehensive books and reports focused in it are rare. I am happy to inform you that the book 'Home gardening' has been compiled on the basis of scientific research works. It will render technical support both on vegetables production, protection, management, consumption, and capacity building activities. I am sure that the information given this publication will be of immense value for those engaged in any form of home gardening and small scale vegetable production as well as training activates.

(M.S. Swaminathan)

Preface

Home gardening is mixed cropping of fruits, vegetables, herbs, spices and other useful plants as a supplementary source of food and income. It is an age-old practice to supply a diverse range of fruits and vegetables to home, but its potential has yet to be fully exploited. Targeting interventions to optimize home garden production and consumption practices show great potential to reduce malnutrition. Promoting home gardening directly or indirectly with different objectives will supplement better and improved access to diversified nutrient packed fruits and vegetables for improved nutrition of household members. There are important gaps in our knowledge about the optimal design of home garden interventions and approach in this important domain. Further there is need to think vertical, not just horizontal as 30 % of Indian rural population is functionally landless.

World population will increase to 8.1 and 10 billion by 2020 and 2050, respectively. The situation is further aggravated by rapidly decreasing availability of land for food production. Home garden is an ancient form of agriculture and an integral component of a wider community development. In its various forms, it remains today one of the most efficient sources of nutrition and income for poor families. Successful interventions are necessary for effective information, education and communication methods to promote home gardening among rural and urban people. There is growing interest in the potential of home garden interventions to address micronutrient undernutrition in low income countries, but evidence is lacking for sustainable impact of such interventions.

South Asia is home to a large proportion of the world's poor, most of whom live in rural areas. As poverty increases, diets often become less diverse. Vegetable production in South Asia has been virtually static for the last two decades and per capita availability has declined as populations have increased. India has emerged as a major producer of horticultural crops surpassing food grain production first time in 2013-14, the trend continued in 2014-15 (283.5MT) and India is placed second after China in both fruit and vegetable production.Peri urban and urban horticulture has emerged as a new area for cultivation of horticultural crops grown for consumption and ornamental use within and in the surroundings of urban areas.Promotion of local crops, home gardens even in urban areas with emphasis on micronutrient dense varieties is needed.

Malnutrition in India, is an emergency therefore there is need for immediate attention if the country has to have inclusive growth and development. India has the highest incidence of under-nutrition in the world. Almost 50% of children under 5 are under weight and stunted. Over 30% of adults are under nourished. A lack of essential vitamins and minerals (particularly vitamin A, zinc, iron, folic acid and iodine) often goes unnoticed by affected people and is, therefore called hidden hunger. Fruit and vegetables are an enormous storehouse of active chemical compounds and considered as the cheapest and most easily available sources of carbohydrates, fiber, proteins, vitamins, minerals and amino acids.

Despite India's rapid economic growth, there are still 300 million people living below the poverty line and 75% of the poor live in rural areas. Per capita consumption of vegetables in India ranges from 86 to 200 g/day, compared with FAO's recommendation of 300 g/day. To improve the nutrition of rural families, vegetables production must be increased and better integrated into the region's predominantly cereal-based farming systems. At the same time, the increasing demand for food from the urban poor living in megacities needs to be satisfied.

Based on 2030 projections, further increase in vegetable production will be needed to meet the demand for fresh, export and processing under changing food scenario.

Most of the supply increase could be achieved through higher per-unit productivity and reduction in post-harvest losses, but production also needs to be boosted in non-traditional areas, and in home and village gardens.There are, therefore, good reasons for home garden interventions to actively promote women empowerment.

This home garden book illustrates the wide array of vegetable based approaches and cuisines to overcome micronutrient malnutrition.It has attempted to bring together elements of home gardening to familiarize the reader with the current approaches of growing and utilization of year round vegetable crops.

The present book has been written in the hope that it will create more interest in home gardening and will work as a guide for those who intend to take up vegetable cultivation on a small scale. It covers a wide range of topic from importance to postharvest handling, to give readers a complete understanding of modern home gardening.

The chapters are organized into eight sections (historical and regional perspectives). All the chapters were rigorously peer-reviewed.We thank the reviewers for their insightful comments and critical suggestions. Special thanks to Ms. Shashi Verma for her excellent support in going through the manuscript and offering valuable suggestions.

I hope that this publication will be a humble contribution to all forms of home gardening and will also contribute to alleviation of hunger and malnutrition.

M.L. Chadha
K.L. Chadha
Roohani Pal
Satish Kumar Sain

Contents

1

Introduction to Home Garden

1.1 Introduction

World population is estimated to be about 7.4 billion; expected to reach nearly 9.1 billion whereas India population is likely to range from 1.4 to 1.5 billion or more by the year 2050. To meet the demand, the food production will need to rise by 70%, and it must double in the developing world. South Asia struggles with food insecurity, poverty, and a very high incidence of malnutrition. Malnutrition not only results in increased mortality and health problems including infectious diseases, mental retardation and blindness, but is also responsible for loss of human capital and work productivity. Causes for malnutrition include inadequate food supply, poor feeding practices and limited knowledge about of the importance of vegetables and fruits in tackling this problem. Most of the population subsists on diet based on staple plants ignoring the vitamin-rich foods such as vegetables, particularly green vegetables. Vegetables provide food diversity and are most affordable sources of micronutrients and health promoting phytochemicals. Per capita consumption of vegetables in South Asia remains low (73 kg/year) compared with the recommended 200 kg/year and more recently 400g/day/person, recommended to be taken as five servings of fruits and vegetables, each serving of 80g. Thus, promotion of home gardens for improved diet and nutritional well-being is an easy way to address the health issues indicted above.

1.2 Characteristics of Home Gardens?

Gardens have been established next to homes since prehistoric times. In the villages also, the small area surrounding a house provides good conditions for a garden. The most important characteristics of home gardens is their location adjacent to homes, close association with family activities and a wide diversity of crop species to meet family needs. They have played a central role in household security for food, fuel, and fibre. Diversity in size, form and function make it difficult to define home gardens, but their place in the farming systems of the rural landscape is readily recognized.

A home garden can be defined as a farming system that combines physical, social and economic functions on the area of land around the family home. Home gardens are developed to supply nutritious food all year round including food plants. To generate income from the sale of garden produce; sales and value-adding can contribute substantially to a family's income.To provide a healthy, comfortable and beautiful environment, a productive home garden can contribute to safe recycling and management of household wastes through composting. Gardens offer privacy, shade and flowers for a family and visitors to enjoy.

1.3 Types of Home Gardens

Home gardens worldwide may be divided broadly into **traditional gardens**, resulting from a long history of adaptation of plants to local needs and conditions, and **model gardens**, often developed with external support, ideas and imported technologies. In urban areas and isolated rural areas, a traditional **kitchen garden** may be inexpensively established - a small plot from which vegetables and garnishes are taken each day to improve meals. **Agroforestry gardens** maximize use of scarce land by cultivating crops in multiple layers -trees, vines, understory and root crops. In the current economic climate, home gardens tend to be most important for the poor and people vulnerable to food insecurity. It is interesting to note that in the current economic climate, home gardens tend to be most important for the poor and people vulnerable to food insecurity.

1.4 Traditional Home Gardens in Rural India

Fig. 1.1: Traditional Home Garden in Rural India

1.4.1 Asian Gardens

Asian gardens (Fig. 1.2) provide households with a number of benefits like: preservation of aesthetic and cultural values; production for family nutrition; are the largest single source of household income in many cases; income peaks in non-harvest seasons, when garden serves as both income and food reserve. In Indonesia, gardens are managed more intensely by poor farmers.

In Nepal and Bhutan, spices and medicinal plants are important and wild vegetables regularly supplement home consumption. In Vietnam and parts of China the vegetable-animal-fishpond garden relies on recycling residues.

Fig. 1.2: Home Gardens in Different Asian Countries

1.4.2 African Gardens

These gardens are multi-storied and diverse in humid areas, becoming less complex and diverse where rainfall declines and is less predictable. In very densely populated settlements, gardens are simpler and smaller- a few fruit trees and vegetables such as papaya, lemon amaranth and okra. Gardens are also a strategic insurance against total crop failure from drought or disease.

1.4.3 Latin American Gardens

The Latin American gardens evolved from a range of ethnic influences from pre-Colombian times are still important for subsistence and income generation. Gardens typically contain root crops, spices, herbs, fruit, vegetables and ceremonial or ornamental plants.

1.4.4 Urban Gardens

Urban gardens (Fig. 1.3) have evolved rapidly with increasing urbanization. These gardens are found wherever the minimum gardening requirements are satisfied: near houses, on high-rise apartment balconies, along drains and roads and in temporarily vacant plots (Fig. 1.3).

Fig. 1.3: A View of an Urban Garden

How can home gardens contribute to food supplies, rural employment and incomes in the future? Among other factors, the growing population has led to increasing poverty and insecure food supplies. Poverty in rural populations, combined with undeveloped transport and food production systems, restricts them largely to locally grown products. The role and contribution of home gardens in addressing these problems have been recognized by development organizations since the 1970s, largely as a result of research into farming systems that resulted in greater understanding of farmers and their households in agriculture and rural development organizations.

In order to improve rural and peri-urban livelihoods in developing countries, development organizations have promoted home gardening with one or more of the following objectives:

- to reduce poverty;
- to diversify income and rural employment;

- to improve the quality and quantity of household food supply and improve nutrition;
- to reduce pressure on wild food resources
- to conserve the indigenous plant biodiversity
- to improve the status of women;
- to improve water and waste management at household and community levels;
- to improve the knowledge of school going kids on healthy diet and importance of vegetables in their life

Home gardens are agro-ecosystems located close to the area that serves as a permanent or temporary residence. Within a very small area one can find a combination of fruit trees, vegetables, root crops, grasses, shrubs, ornamentals and herbs that provide food, spices, medicines, flowers and construction materials.

It is an intensive type of vegetable growing to minimize buying from the market providing continuous supply of fresh nutritious vegetables for the family. It includes use of low-cost inputs and indigenous varieties; reducing dependence on exotic or imported varieties and management by members of household (wife, husband, children).

1.5 Options for Home Gardens

Home garden can be in a small piece of land, generally back yard sometimes front yard.(Fig. 1.4)

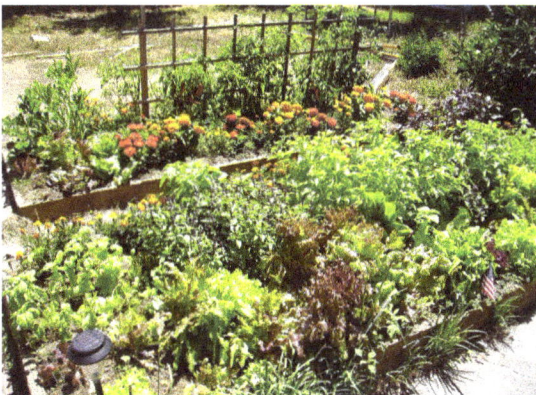

Fig. 1.4

-on the verandah: using special growing containers- plastic pots, plastic bags, clay pots or other convenient containers for growing plants.(Fig 1.5)

Fig. 1.5

-on roof tops: with soil placed on the cemented floor and other containers as in verandah.(Fig 1.6)

Fig. 1.6

-sometimes locating little far away from home: depending upon the availability of land, water and other facilities.(Fig 1.7)

Fig. 1.7

-on special structures: meant for urban localities.(Fig 1.8)

Fig 1.8

-using small space, no soil: hydroponics (Fig 1.9)

Fig. 1.9

Families with home gardens are at a lesser risk of xerophthalmia, night blindness and other problems associated with Vitamin A deficiency due to increase in their daily consumption of safe vitamin A rich vegetables. Home gardens are most suitable for fallow and upland areas where growing other crops poses problems.

1.6 Importance of Home Gardens

1.6.1 For Fresh and Healthy Food

Home gardens are important for sustaining agro-biodiversity and contribute to securing livelihoods and incomes. Families with gardens will eat more healthy fruits and vegetables simply because of their ready availability. Own home gardening also reduces cash outlay for food and provide additional income to farmers. It can be a source of hard-to-find/uncommon and fresh vegetables such as kohlrabi, Chinese cabbage, pakchoi etc. Since, fruits and vegetables grown in one's own garden have not to be shipped for long distances, one is able to receive the maximum amount of nutrients that they can provide. Vegetables from the supermarket also do not compare in taste, quality, or freshness with vegetables grown in the home garden. Avoiding the use of artificial chemicals and pesticides in one's home garden is a step to ensure that food is healthy and safe for the family to eat. Excess produce of fruits and vegetables from the home garden can be frozen, canned, or dried so that they can be enjoyed year long.

1.6.2 For Physical Activity

For many people raising vegetables in home garden is a pleasant way of exercising and economizing. Gardening can be a great, low-impact exercise. Doctors suggest 30-60 minutes of low to moderate intensity physical activity per day in order to maintain a healthy heart and quality weight.

Table 1.1: Comparison of Energy Expenditure of Gardening and Wellness Center Equipment

	Energy Expenditure (kcal)
Total EP for a 6 x 6 m2 gardening activities	**53786**
Working in the garden	**Energy expenditure (kcal/hour)**
One hour/day for 3 months	598
Two hours/day for 2 months	448
Two hours/day for 3 months	299
Using the wellness center exercise equipment	
Treadmill (speed 6.0km/hour)	270
Treadmill (speed 3.0km/hour)	168
Treadmill(speed1.5km/hour)	120
Elliptical trainer (speed 10.7km/hour)	300
Recumbent bike (speed 1.2km/hour)	240

Typical calories burnt (in adults) during 30 minutes of activity in home garden are: hauling heavy rocks – 300, mowing (push mower) – 243, digging, spading, tilling – 202, mowing (push mower with motor) – 182, trimming shrubs – 182, planting trees – 182, weeding – 182, raking – 162, planting seedlings – 162, mowing the lawn (riding) – 101 and watering the lawn or garden – 61.

1.6.3 For Mental, Emotional, and Social Well-Being

For many people, the garden is an escape from stress and an ideal place for relaxation. It is the cheapest, healthiest, keenest pleasure one can get. It provides an opportunity for the family to work together and strengthen ties.

1.6.4 Economic Role of Vegetables

vegetable production is labor intensive, creates job opportunities in the rural and sub-urban areas, such as marketing, processing and transportation. Women play a role in vegetable production. They not only tend the gardens but harvest, sell the produce and cook meals.

1.7 Types and Designs

Cities have limited land therefore, vegetables have to be grown in a compact area or in the containers, while in the villages land may not be a limiting factor but choice of seeds types of vegetables in the garden may be limited. Between village traditions, eating habits and availability of resources like water and other inputs in and around different villages determine the types and designs of home garden.

Some families may have choice of a particular vegetable more where as other families may have different choices. All these choices determine the designs and types of garden. One may decide to plant more of a particular variety to satisfy their taste regularly; where as other family may do other way round. These choices determine the designs and types of garden.

Family size also makes a difference for a large size family may need a bigger garden than a small size family, hence design too has to be for the has to be different.

1.8 Importance of Vegetables

- High intake of vegetables as part of a healthy diet can make an important contribution in preventing chronic non-communicable diseases and their risk factors.
- Dietary diversification induced through higher vegetable consumption balances the diet by enhancing the supply of essential micronutrients leading to improved health, enhanced cognitive ability, and increased efficiency
- Enhances vegetable cultivation by naturally enriching soils, diversifying cropping systems, and reducing pest populations.
- Vegetables provides an opportunity to the farmers for earning extra income in short duration by cultivating vegetables in between the main crops or as an alternate to the main crops.

- Develops management and leadership skills among homegardners.
- Processing vegetable crops for value addition creates additional job opportunities, especially for women in the rural sector.
- Vegetable production supports creating new economic and entrepreneurial opportunities, especially for women.
- Help combat malnutrition, diversify diets, and alleviate poverty by creating jobs and new sources of income for smallholder farmers and landless laborers.

Gardening is a fun, and when you can make a direct impact on your health, it makes home gardening a lot more purposeful. Have a great garden filled with natural ways to treat. For many people, raising vegetables in home garden is a pleasant way of exercising and economizing on food cost.

2

Planning

2.1 Vegetable Crop Families and Crop Selection

It can be hard to know which vegetable belongs to which plantfamily, just by looking at it, but understanding the major plant families would make the task a little less daunting. Most home vegetable gardeners grow several plant families in any given year ,thus by using a handy vegetable families list would help planning easy.

2.1.1 Solanaceae

The nightshade family is perhaps the most commonly represented group in most home gardens. Members of this family include tomatoes, peppers (sweet and hot), eggplants, potatoes (but not sweet- potatoes). *Verticillium* and *Fusarium* wilts are common fungi which build in the soil when nightshade members are planted in the same spot year after year.

2.1.2 Cucurbitaceae

The vines of the gourd family or cucurbits may not seem similar enough to be closely related at a first glance, but each and every member produces fruits on a long vine with seeds running through the centre,and most of them are protected with a hard rind. Cucumber, zucchini, summer and winter squash, pumpkin, melonand gourd are members of this large family.

2.1.3 Fabaceae

This family of legumes is a large family; important to many gardeners as nitrogen-fixers. Peas, beans, peanuts and cowpea are common vegetables in the family.

2.1.4 Brassicacae

Also known as cole crops, members of the mustard family tend to be cool-season plants and are used by many gardeners to extend the growing season. Some gardeners say that the flavour of thick-leafed members of this family improve a little by frost. Broccoli, cauliflower, cabbage, kale, Brussels' sprout, radish, turnip and other mustards are grown in home gardens.

2.1.5 Liliaceae

Not every gardener keeps space for onions, garlic, chive, shallots onion, but if done, these members of the onion family require rotation just like other families. Although asparagus may be grown in a place for several years, and when selecting a new site for asparagus beds, make sure that no other family members have grown nearby for several years.

2.1.6 Lamiaceae

Not technically vegetables, many gardens may contain members of this mint family, which benefits in crop rotation owing to several persistent and aggressive soil-borne fungal pathogens. Members such as mint, basil, rosemary andthyme are sometimes inter-planted with vegetables to deter pests.

2.2 Selection of Vegetables and Herbs for Home Garden

It is commonly done by what appeals to senses of taste and smell. An individual crop's requirement for spacing, sun preference, watering, fertility and upkeep may causse limitations on home gardener's ability to grow or care for that crop.

Fortunately, different varieties of vegetables enable gardeners togrow many types of crops in different types of spaces and situations. Breeders are bringing more varieties to home garden than ever before. Whether a gardener's limitations are dictated by a lack of space, disease resistance or length of growing season, the right plant to fit in a particular niche has never been so accessible as at present. The following are some common vegetables and herbs.

Vegetables

Common Name	Botanical Name
Beans	*Phaseolus vulgaris*
Beet	*Beta vulgaris*
Broccoli	*Brassica oleracea*
Brussels Sprouts	*Brassica oleracea*
Cabbage	*Brassica oleracea*
Cantaloupe	*Cucumis melo*
Carrot	*Daucus carota*
Cauliflower	*Brassica oleracea*
Celery	*Apium graveolens*
Chicory	*Cichonum intybus*
Chinese Cabbage	*Brassica rapa*
Collards	*Brassica oleracea*
Corn	*Zea mays*
Cucumber	*Cucumis sativus*
Eggplant	*Solanum melongena*
Endive	*Cichonum endvia*
Kale	*Brassica oleracea*
Kohl Rabi	Brassica oleracea
Leek	*Alliumampeloprasum*
Lettuce	*Lactuca sativa*
Mustard	*Brassica jncea*
Okra	*Abelmoschus esculentus*
Onion	*Allium cepa*
Pak Choi	*Brassica rapa*
Parsnip	*Pastinaca sativa*
Peas	*Lathyrus odoratus*
Pepper	*Capsicum annuum*

Contd...

Common Name	Botanical Name
Pumpkin	*Cucurbita*
Radish	*Raphanus sativus*
Rutabaga	*Brassica napobrassica*
Soybean	*Glycine max*
Spinach	*Spinacia oleracea*
Squash	*Zucchini*
Swiss Chard	*Btea vulgaris*
Tomato	*Lycopersicum*
Turnip	*Brassica rapa*
Watermelon	*Citrullus lanatus*
Wheatgrass	*Triticum aestivum*

Herbs

Common Name	Botanical Name
Anise	*Pimpinella anisum*
Balm Lemon	*Melissa officinalis*
Basil	*Ocimum basilicum*
Buckwheat	*Fagopyrum esculentum*
Catmint	*Nepata mussini*
Catnip	*Nepeta cataria*
Chervil	*Anthriscus cerefolium*
Chives	*Allium schoenoprasum*
Coriander/Cilantro	*Coriandrum sativum*
Clary	*Salvia sclarea*
Cumin	*Cuminum cyminum*
Dill	*Anethum graveolens*
Fennel	*Foeniculum vulgare* var *dulce*
Hyssop	*Hyssopus officinalis*
Lavender	*Lavendula vera*

Contd...

Common Name	Botanical Name
Lovage	*Leristicum officinalis*
Marjoram	*Marjorana hortensis*
Mint	*Mintha spicata*
Oregano	*Origanum vulgare*
Pak Choi, Ming Choi	*Cinakohl pak choi*
Parsley	*Petroselinum crispum*
Rosemary	*Rosemarinus officinalis*
Savory	*Satureja hortensis*
Sorrel	*Rumex acetosa*
Stevia	*Stevia rebaudiana*
Thyme	*Thymus vulgaris*
Watercress	*Nasturtium officinalis*

2.3 Twarm-Season Vegetables

The following vegetables are considered ideal for the warm season.

2.3.1 Corn

Most corns are best in hot-summer areas, but early-maturing hybrid varieties would grow even in regions with cool summers. Sow seeds directly in the garden, spacing tat 4 - 6 inches apart in rows distanced 30 to 36 inches . Thin seedlings at1- 1½ feet apart. Harvest 60 to 100 days after sowing.

2.3.2 Cucumber

Cucumbers are easy to grow from seed; sow two or three seeds in groups spaced 1½ feet apart at the base of a trellis; then thin seedlings. Harvesting is started 50 to 100 days after sowing. Cucumbers thrive in the heat of summer and their trailing vines can sprawl across the garden. Bush types are excellent for small-spaced gardens and containers. Cucumbers are one of the most preferred vegetables all around the world; they taste great and are cool and refreshing.

2.3.3 Eggplant

It is a small- to medium-sized bush vegetable producing
smooth, glossy skinned fruits ,which can vary in length from
5 to 12 inches. The edible fruit can be long and slender or
round or egg-shaped. Fruit is creamy-white, yellow, brown,
purple, or sometimes almost black. Eggplants can grow 2
to 6 feet tall, depending on the variety. Grow eggplant in
full sun and it will grow best in well-drained soil rich in organic matter;it is sensitive
to cold. It grows best where day temperatures are between 27° and 32°C and night
temperatures between 21° and 27°C.

2.3.4 Hot Peppers

They likewise offer a range of sizes, colors and
pungenciesThe heat of hot peppers intensifies as pe ppers
ripen. They are of a wide range of colours from yellow,
orange, purple, and even brown on ripening. Some chili
peppers turn bright red, which more often is an indication
of ripeness than hotness. One plant will produce many
during a growing season,

2.3.5 Melons

Cantaloupes (also known as muskmelons) are the
easiest melons to grow, as they ripen the fastest. Sow four or
five seeds per hill; space hills 48 - 72 inches apart. Thinning
of seedlings is done to two per hill. Harvest 70 to 115 days
after sowing.

2.3.6 Snap Beans

They are also called string or green beans and have tender, fleshy pods. Besides
the familiar green types, yellow or purple pods are also available. One can choose
self-supporting (bush) or climbing (pole) varieties. Plant seeds of bush types, 2 inches
apart, in rows spaced at 24 -36 inches; thin seedlings at 4 inches distance. For pole

beans, space seeds at 4 - 6 inches and distance among rows should be 36 inches; support plants on trellis. Thin seedlings at 6 inches. Begin harvesting 50 to 70 days after sowing.

2.3.7 Squash

There are two basic types of squash. Summer squashes (zucchini, crookneck, pattypan) are eaten when the fruit is small and tender; harvested 50 to 60 days after sowing. Winter squashes form hard shells; they are harvested in fall (80 to 120 days after sowing) ,and can be stored for winter season. Sow seeds of bush types 12 inches apart in rows at a distance of 36 to 60 inches; thin seedlings to 24 inches apart. Sow seeds of vining squash in hills spaced 5 feet apart, placing four or five seeds in each hill; thin them to two per hill.

2.3.8 Sweet- pepper

They come in all shapes and sizes. Left to ripen, they turn red, purple, orange, or yellow and have varied amount of sugar depending on the variety. Green bell peppers are the most common ones. Because peppers require a long, hot- growing season, in cool regions or areas with short- growing seasons, they may never develop their ripe color. Start sowing seeds of sweet or hot peppers in indoors 6 to 8 weeks before planting time; or buy transplants. Set plants out; spacing them 18 to 24 inches apart in rows 2½ feet apart. Harvest 60 to 95 days after setting out plants.

2.3.9 Tomato

Easy to grow and prolific, tomatoes are a homegarden favourite. A huge number of varieties are available, varying from tiny cherry types to giants; fruit colours include red, yellow, orange, and even pink. Sow seeds in flat bed 6 weeks before planting time; or buy transplants. Set out in the garden by spacing plants 24 to 48 inches apart in rows distanced 3 to 4 feet apart. Bury as much as half to three-quarters of the stem of each plant; roots will form along the buried part and strengthen the plant. Stake plants or place wire cylinders around them for support.

2.3.10 Zucchini

Another fastest growing vegetable from seeds is zucchini. Zucchinis are favourite vegetables in most countries, and they can grow well in the garden if provided with basic inputs. Zucchini can grow within 70 days and can be harvested with ease.

2.4 Cool-Season Vegetables

2.4.1 Beans

They are one among the easiest vegetables to grow, and is perfect for a home garden; beans germinate quickly and produce copious amount of tasty treat. They are available in a variety of shapes, colours and sizes; some plants produce colourful flowers, pods and seeds. From snap beans to edamame (vegetable soybean)—you can grow them all. Beans are easy to grow from seeds.

2.4.2 Beet

In beets,besides basic red, golden yellow and white varieties are also available. The tender young leaves are also edible. Plant seeds 1 inch apart in rows spaced 18 inches apart, or broadcast them in wide beds; thin seedlings to 2 to 3 inches apart. Harvest 45 to 65 days after sowing. Beetroots often used in salads but is equally tasty when eaten warm and freshly boiled vegetable

2.4.3 Broad Beans

What could be simpler! Sow broad beans and within a few weeks these growing beans would make sturdy plants.

2.4.4 Broccoli

Easy-to-grow broccoli bears over a long season. Start seeding 6 weeks before planting time; or buy transplants, set out plants 15 to 24 inches apart in rows spaced 2 to 3 feet apart. Seeds can be sown directly in the garden, spacing them 4 inches apart; thin seedlings to 15 to 24 inches apart. Harvest 50 to 100 days after setting out plants,and 90 to 140 days after sowing. Cut heads before buds begin

to open. After the central head is harvested, side shoots would produce additional smaller heads.

2.4.4 Cabbage

Sow seeds 6 weeks before planting time; or fetch transplants, set plants 15 to 24 inches apart in rows spaced 24 to 48 inches apart. Seeds can be sown directly in the garden, spacing 4 inches apart; thin seedlings to 15 to 24 inches apart. Harvest 50 to 100 days after setting out plants;90 to 140 days after sowing.

2.4.13 Carrot

Sow carrot seeds half to one inch apart in rows spaced 1 to 2 feet apart; or broadcast seeds in wide beds; thin seedlings at2 to 4 inches apart. Harvest baby carrots 30 to 40 days after sowing and mature carrots 50 to 80 days after sowing. Keep the soil moist, but not soaked. Expect the seeds to be germinated in about two weeks. Carrots are ready to harvest when they have grown to about ¾ of an inch across the top (just below the green stem).

2.4.3 Cauliflower

It requires consistently cool temperature ; sow seeds ½ to ¾ inches deep. Rows should be three to six inches apart, with a maximum of eight seeds per 12 inches of row. Transplantation of cauliflower is on 8- to 10-inch rows at least 36 inches apart. When wateredfor the firsttime, use a high phosphate fertilizer so that plants can have a good contact with soil;water every five to seven days. This is required for to get nice heads. Harvest cauliflower, know that mature heads are fully developed, clear, white and compact head 6 inches in diameter. Cut the head from the plant with a large knife, leaving at least one set of leaves to protect head.

2.4.12 Green Onion

This is one of the fast- growing plants. Green onion stalks can be easily harvested after three to four weeks. Growing normal onions may take around 6 months or more than that.

2.4.6 Green Peas

They are trouble -free and can grow better in cooler weather. Peas grow pretty fast when planted in a home garden. Once peas are sown in the soil, they would germinate within ten days. After 60 days, fully grown ready to be harvested pods would appear .

Some peas are for shelling, some have edible pods; and some can be harvested either way. Bush and vining types are available. Sow seeds 1 inch apart in rows spaced at 24 to 36 inches. Thinning of seedlings should be done 2 to 4 inches apart. Set up stakes or trellises for vining types at planting time. Harvest 55 to 70 days after sowing.

2.4.5 Lettuce

Sow seed directly in the garden or set out transplants; make successive plantings or sowings until daytime temperature reaches between 24° and 27°C. Sow lettuce seeds 1 to 2 inches apart in rows spaced at 12 to 24 inches apart; thin to 6 to 8 inches apart. Seeds can be broadcast in wide beds; thin to 6 inches apart. Harvest leaf lettuces 40 to 50 days after sowing; and some late varieties in 65 to 90 days.

2.4.11 Onion and Garlic

These crops are virtually maintenance free and are really such easy vegetables to be grown! Simply plant onion bulbs and individual garlic cloves on well-drained soils, then leave them! In summer when the foliage turns yellowand dieback, they can be lifted and dried in the sun before storing.

2.4.14 Radish

They are very easy to grow in home garden, just sow seeds, mark the area and water them. Within 25 to 30 days one would be able to harvest radishes.

2.4.7 Snow Peas

They thrive in cool, moist weather, and are eaten fresh. Some peas convert as much as 40 % of their sugar to starch in just a few hours in refrigerator. Snow peas, and their close relative snap peas, are eaten as a whole; thus no time-consuming shelling is required.

2.4.8 Spinach

It bolts quickly into flower if the weather gets too warm or the day is too long. For best results, sow seeds 1 inch apart in rows distanced 12 to 30inches; or broadcast seeds over wide beds. Thin seedlings to 3 to 4 inches apart. Harvest 40 to 50 days after sowing. Spinach is easy to grow from seeds.

2.4.15 Turnip

Its seedsare sown in the soil about 1/2 inch deep at a rate of three to 20 seeds per foot. Water them immediately for speedy germination. Once turnip seedlings emerge, thin them about 4 inches apart to give plants plenty of room to form good roots. Plant turnips at the interval of ten days, which would allow harvesting every couple of weeks throughout the season. Harvest in about 45 to 50 days after planting.

2.5 Selected Herbs

2.5.1 Basil

It is one of the easiest herbs to grow. Add it to sauces, soups and salads for a spicy, tangy flavour. Many varieties, from lime basil to Thai basil, have flavours ranging from citrusy to spicy. Basil grows equally well in the garden and in containers, and its clean, long-lasting foliage makes it a great edible plant. Basil is easy to grow from seed or transplants. Its leaves have warm, spicy flavour.

2.5.2 Chives

Sow seeds ½ inch deep in rows 12 inches apart. As soon as the seedlings are established, thin them in rows 6 inches apart. Or set out nursery grown plants 9-12 inches apart. Leaves have a mild onion flavour. Chop them and add them to salads, egg and cheese dishes, mashed potatoes, sandwich spreads and sauces. Use flowers in salads.

2.5.3 Coriander

Sow seeds, ¼ inch deep in rows 12 inches apart. Thin established seedlings to 6 inches apart. Grind dry seeds to powder and dust over non-vegetarian and vegetarian dishes before cooking. Young leaves are knows as cilantro. Roots, which can be frozen, are used to flavour soups.

2.5.4 Mint

It is so vigorous that it will grow on almost any moist soil; so it is best to keep it in a pot to stop its spread too far,Plant 4- to 6-inch pieces of root 2 inches deep and 12 inches apart. Water well. Check roots' tendency to overtake nearby plant roots by sinking boards or bricks 1 foot deep around beds

2.5.5 Parsley

Soak seeds overnight and broadcast thinly. Thin the established seedlings to 9-10 inches apart. Mix leaves into salads, soups, stews and omelets. Serve fresh as garnish to meat, fish, and onion dishes

2.5.6 Rosemary

The heavenly-scented herb is an antioxidant that may help limit weight gain and improve cholesterol levels. Start by planting seeds (or propagating cuttings) in a sunny area of the home garden; rosemary will grow best with at least six hours of direct sunlight each day. **Water only when the top of the soil is dry,** be sure not to let the soil dry out completely. Gently snip a few sprigs from each plant, being sure not to remove all of the leaves from any one plant.

2.6 Factors in Selecting the Garden Site

Following factors need to be considered: Sunlight, Nearness to the house, Soil, Water, Air drainage and Shelter.

2.6.1 Sunlight

All vegetables need defined period of sunlight. The garden should receive at least 6 hours of direct sunlight each day. Every day 8 to 10 hours sunlight is ideal. To ensure this, vegetables should be planted away from the shade of buildings, trees and shrubs. Plots sloping a little to south or east seem to catch sunshine early and hold it late, and that seem to be out of the direct path of chilling north and north-east winds are suitable. Some leafy vegetables such as broccoli, spinach, mint and lettuce tolerate shadier conditions than other vegetables. If a shady site is the only choice, leafy crops and root crops would do better than flowering and fruit crops such as squash, cucumber, pepper and tomato.

2.6.2 Nearness to the House

The notion of a "kitchen garden" is popular because it is easy to walk out your kitchen and pull fresh produce and herbs. The closer is the vegetable garden, easier and more frequent is to use it. Harvest vegetables at their peak and take maximum advantages of garden freshness. It would also be more convenient to take up operations like weeding, watering, insect and disease control, and succession planting if the garden is closer by.

2.6.3 Soil

The soil should be fertile and easy to till with just the right texture a rich, sandy, light, well-drained loam. **Rich** means full of plant food, ready to be used at once, all prepared and spread out on the garden table. Soils can be made rich or kept rich in two ways; first, by cultivation, which helps change raw plant food stored in the soil into available forms; and second, by manuring or adding plant food to soil from outside sources. **Sandy** means a soil containing enough particles of sand so that water passes through it without leaving it pasty and sticky a few days after the rain. **Light**, so much that a handful of soil, under ordinary conditions, will crumble and fall apart readily after being pressed in hand. **Loam** is a rich, friable soil, in which sand and clay are in proper proportion, neither predominating, and is usually dark in colour from cultivation and enrichment. A dark, loamy soil enriched with plenty of organic matter warms up more rapidly and retains heat better than either sandy or heavy clay soils. Avoid any soil that remains soggy after a rain. Heavy clay and sandy soils can be improved by adding organic matter. Gardening would be easier with a naturally rich soil.

2.6.4 Water

Water is vital, right from the moment seeds are sown through sprouting to end of the growing season. Plants need water for cell division, cell enlargement, and even for holding themselves up. If the cells don't have enough water, the result is a wilted plant. Ideally, water for plants comes from rains or other precipitation and from underground sources. But, we often have to do extra watering on need basis by hand or through an irrigation system. Therefore, it is essential to have location of the garden near a water source. The best time to water garden is in the morning. If watering is done at night when the day is cooling off, it is likely to stay on the foliage, increasing danger of disease incidence and spread.

When to water home garden: Always soak the soil thoroughly. A light sprinkling can often do more harm than no water at all. It stimulates roots to come to the surface, where they are killed by exposure to sun. More water evaporates when temperature is high than when it is low. Plants can rot if get too much water in cool weather. More water evaporates when relative humidity is low. Plants need more water when the days are bright. Wind and air movement increases loss of water to atmosphere.

Water needs vary with the type and maturity of the plant. Some vegetables are tolerant to low soil moisture. If the day is hot and the plants wilt in the afternoon, don't worry about them; they will regain their balance overnight. But if plants are wilting early in the morning, water them immediately.

2.6.5 Air Drainage

Avoid locating garden in a low spot such as the base of a hill or the foot of a slope bordered by a solid fence. These areas are slow to warm in the spring, and frost forms more readily in them because cold air cannot drain away. Home gardens located on high ground are more likely to escape light freezes, permitting an earlier start in the spring and a longer harvest in the fall.

2.6.6 Shelter

Most crops can be severely affected in windy sites. Even light winds may reduce yields by 25%. If the site is exposed, set- up windbreaks to baffle and filter the wind. A windbreak is a hedgeor afence. Locating the garden within a windbreak improves yield and quality of its contents. Lower wind speeds reduce damage to tomato, pepper, leaf lettuce, pea, beans and other garden vegetables. And bee activity increases, resulting in complete pollination and fruit formation. The moderation of the microclimate by the windbreaks results in early maturity of crops such as asparagus, tomato, sweet- corn, cucumber and melons. **Windbreaks** capacity to block flow of wind depends upon their height and density or thickness. A home garden windbreak also can serve as a fence.

Apart from preventing loss of fertile soil, windbreaks also reduce loss of soil moisture and improve microclimate. Some common windbreak plants are: *Azadirachta indica, Bixaorelana, Cassia, Casuarina equisetifolia, Erythrina, Leucaena leucocephala, Parkinsonia aculeata, Prosopis juliflora, Schinusmollis, Tamarix.*

2.7 Soil Management

The soil is the most easily controllable part of the plant's environment. It provides mineral nutrients and water used by plants to manufacture food supply and structural components. Also it supplies oxygen and structural support for plant roots. Soil fertility can be defined as the quality of the soil enabling it to supply proper kind and amount of elements needed for vegetables growth when other factors such as light, temperatureand other soil characteristics are favourable.

2.7.1 Soil as a Plant-Nutrient Medium

Physical, chemical, and biological soil properties are important to plant nutrition. The soil can be said to have three major components: the **solid phase**, the **liquid phase** and the **gaseous phase**.

i) Solid phase: It is the main nutrient source. It holds the primary minerals containing essential macro and micro-elements and large organic particles, mostly made up of C, H, O, and smaller amounts of other essential elements. It also has the finer (colloidal) particles where exchangeable cations are adsorbed.

ii) Liquid phase: In this, nutrient elements occur in ionic forms to be readily absorbed by plant roots.

iii) Gaseous phase: It is where gas exchanges between living organisms in the soil and the atmosphere take place.

The soil, therefore, is not only a nutrient reservoir but also a medium through which nutrient elements are made available to plants. To promote soil fertility, adequate supply of nutrients must be maintained in the soil and these nutrients must be available at a rate suitable to normal vegetable growth. An adequate nutrient ratio is necessary; the total concentration being equally vital.

2.7.2 Soil Texture

Soil is composed of three basic mineral particles of three different sizes. Sand is the largest particle, silt is intermediate, and clay is the smallest. The percentage of each determines soil texture as well as its physical properties. An ideal soil texture consists of equal parts of sand, silt, and clay. Such a soil is referred to as "**loam**". Usually, one component predominates, producing a clay- loam or a sandy- loam. To check the texture of the soil, take a moist sample between your fingers and rub it. Sandy soils tend to be harsh and gritty; clay and silt are smooth and somewhat slippery. Another test is to form a ball of moist soil with your hand. If the ball breaks apart when tapped, the soil is said to be on the sandy side. If the ball remains intact when tapped, the soil probably contains more clay and silt than sand. If the soil is either sticky or plastic, and works through your fingers as you form the ball, considerable amount of clay is present.

2.8 Soil Nitrogen and Organic Matter

The forms of soil nitrogen are organic nitrogen, ammonium nitrogen, soluble inorganic nitrate compounds, and ammonium nitrogen, fixed by certain clay minerals. Soil nitrogen is always in a state of change. Nitrogen may be transformed through immobilization, mineralization, nitrification and denitrification.

i)   Immobilization: Microbes decompose plant and animal residues. Through this process, nitrogen becomes tied up in the tissues of microbes.

ii)   Mineralization (ammonification): It is conversion of an element (inorganic combination) to available form as a result of microbial decomposition (breakdown of organic N to inorganic form).

iii) Nitrification: It is a microbial reaction converting ammonium nitrogen to nitrate form (formation of nitrates from ammonia in the soils by soil organisms). Aeration, temperature (approximately 30 °C optimum), moisture, available calcium and magnesium, and fertilizer enhance microbial nitrification.

iv) Denitrification: It is a process where soil organisms, particularly anaerobic organisms, convert nitrate nitrogen to gaseous form, which is then lost from the soil (the reduction of nitrates to nitrites, ammonia and free nitrogen).

When organic residues with a wide carbon-nitrogen (C/N) ratio are added to the soil which contains nitrate nitrogen, decay organisms prevail and nitrifying bacteria become more or less inactive. During the nitrate depression period, plants can have little nitrogen from the soil. Fertilizers, manures and the return of crop residues are major nitrogen sources. Under field conditions, the greatest part of soil nitrogen is present in organic form and is slowly released to plants after conversion to inorganic forms by mineralization and/or nitrification. Nitrogen is lost through harvest and excessive leaching. Fertilizer should be applied properly to ensure farm productivity and to avoid unnecessary loss of elements in the soil.

2.8.1 Management of Soil Nitrogen

An adequate supply of nitrogen can be maintained by adding compost, including legumes in crop rotation, and applying fertilizer nitrogen. However, the soil can only hold a certain amount of nitrogen, beyond which, raising the nitrogen content will only result in unnecessary nitrogen loss by leaching, volatilization and erosion.

2.8.2 Microbial Nitrogen Fixation

Legumes (*Rhizobium* species) fix nitrogen. Additionally bacteria like *Azospirllium, Azotobacter* fix nitrogen in non-leguminous plants, especially cereals and vegetables The amount of nitrogen fixed by the bacteria depends on drainage, moisture, pH, and available calcium. Nitrogen fixation by blue-green algae and *Azolla* is also very significant.

2.9 Soil Phosphorus

Small amount of phosphorus is present in the soil, which is relatively unavailable in the native form,and phosphorus supplied through fertilizer can be fixed in several ways. Inorganic phosphorus (approximately 50% of the total) such as calcium-containing phosphorus compounds are formed especially in high pH; while iron and aluminum compounds are formed at low pH.

2.9.1 Organic Phosphorus

Forms in organic matter include phytins, nucleic acids and phospholipids. Availability of organic phosphorus is determined by the amount of organic matter in the soil and its rate of decomposition and activity of microbes. The fixed phosphorus in the soil can be solubilized by phosphate solubilizing bacteria (PSB), which have the capacity to convert inorganic unavailable phosphorus form to soluble forms (HPO_4^{-2} and $H_2PO_4^{-1}$) through process of organic acid production, chelation and ion -exchange reactions and make them available to plants. More phosphorus is obtained by applying fresh organic materials. Strains from bacterial genera *Pseudomonas, Bacillus, Rhizobium* and *Enterobacter* along with *Penicillium* and *Aspergillus* fungi are the most powerful P solubilizers.

2.9.2 Control of Phosphorus

Maintain soil pH (6-7) and level of organic matter. It is good to use fertilizers covered or coated by various materials. Despite careful management, a large portion of added phosphate remains unavailable to the plant. This phosphorus is not subjected to leaching loss, remains in the soil and can slowly become available to the plant.

2.10 Soil Potassium

Most, except sandy soils, are high in total potassium. Quantity of potassium held in an exchangeable form in clay is very low. Much of potassium is lost through leaching. Potassium and nitrogen removed by the crop are very high. Many vegetable crops have very high potassium requirement. About 90-98% of the soil potassium in the form of feldspars and micas is not available to plants ;only about 2% is available. Of this, potassium in solution makes up only 10% and is subject to leaching. Remaining exchangeable potassium is held in soil- clay.

Factors which affect potassium fixation in the soils depend on the type of clay. Liming may sometimes increase potassium fixation in the soil. Frequent light applications of potassium fertilizers are usually better than large and less frequent ones. As plants take up large quantities of potassium, returning crop residues to soil becomes very important.

2.11 Soil Sulphur

Natural sources of sulphur are incorporated in soil minerals, such as sulphides of iron, nickel, and copper. Atmospheric sulphur from the combustion of coal is carried down to soil by rain. Some organic forms also have cycles of immobilization and availability similar to nitrogen. Deficiencies increase with less sulphur being added and as fertilizers become more pure in nitrogen, phosphorus and potassium.

2.12 Soil Calcium

Calcium is more commonly present in the soil than other plant nutrients. The most common sources of calcium in the soil are minerals like calcite ($CaCO_3$), dolomite [$Ca_2Mg(CO_3)_2$] and gypsum ($CaSO_4.2H_2O$). Calcium is liberated when these minerals disintegrate and decompose. The calcium content of soils varies widely from as low as 0.1% in the humid tropics to as high as 25% in calcareous soils. Available calcium exists in the soil as calcium ion (Ca^{2+}) in the soil solution and as exchangeable ion absorbed on the clay complex. The calcium ion in the solution is easily leached when there is too much water and especially in light sandy soils.

2.13 Soil Magnesium

Its content in the soil varies from 0.1% in the coarse, sandy soils in the humid regions to about 4% in the fine-textured and semi-arid soils. Primary sources of magnesium are such minerals as biotite and dolomite. Magnesium available to plants is in the exchangeable and water-soluble forms of its ion (Mg^{2+}). Its absorption by plants depends on the amount of K, soil *p*H, degree of magnesium saturation of soil colloids, nature of other exchangeable ions, and type of clay – the same factors ,which affect calcium absorption by plants. Moreover, its ions also leach easily.

2.14 Soil Micronutrients

Iron, manganese, zinc, copper, boron, molybdenum, and chlorine are necessary for plant growth. Crop removal reduces amount of trace elements below the level required for normal plant growth. Use of improved crop varieties and macronutrient fertilizers has increased crop production levels tremendously, resulting in greater removal of micronutrients. There is also concern over the level of trace elements in the soil and in the food supply, as most of these elements are also necessary for human nutrition.

Generally, these elements are present in inorganic forms as primary soil minerals and clays, and also in organic forms. Reasons for micronutrient deficiency are leaching, acid and sandy soils, organic soils (histosols), soils with very high pH, soils intensively cropped and soils heavily fertilized with nitrogen, phosphorus and potassium only. Excessive amounts of zinc, manganese, or copper may cause iron deficiency, while heavy fertilization may cause copper deficiency and excessive potassium may decrease manganese and calcium uptake, and *vice versa*. *Pseudomonas* and *Bacillus* have been identified as the potential target organisms and can solubilize zinc sulphide, zinc oxide and zinc carbonate.

2.15 Soil pH

The pH of the soil solution greatly influences many of the important soil and plant processes. Harmful effects do not come from H^+ ions themselves, but from toxicity of Al^{3+} and Mn^{2+} ions which are soluble at low pH. At higher pH, bicarbonate ions may become toxic. Alkaline soils often have high salt concentration, which is harmful to plant growth. Salt concentration is measured by electrical conductivity (EC) of the soil. EC values of more than 4 mS/cm are harmful for plants. Generally, pH between 6.5 and 8.5 is normal for cultivation of vegetables.

2.16 Improvement and Maintenance of Soil Fertility

The physical properties of soil may be improved by plowing, inter-tillage, organic materials application (compost, manure, and green manure), and crop cultivation. There are two basic ways to improve soil chemical properties.

- Application of adequate amount of nitrogen, phosphorus and potassium, organic materials, liming of acidic soils, and application of deficient elements.

- Removal of inhibited elements for plant growth by adjusting soil pH and reducing toxic heavy metals.

- Other methods include: soil-erosion control; crop rotation; soil microbial properties may also be improved by keeping soil free of disease- causing microorganisms.

2.17 Fertilizers and Fertilizer Management

Fertilizers supply a combination of food elements to help in plant growth and development. Plants need certain nutrients from the soil for growig. Sixteen elements are known to be essential for the growth and development of plants. Three of these come from air and water and the rest come from soil. Different vegetables have different requirements of these elements: only some species require all of them. Leafy vegetables need more nitrogen (N), flower and fruit vegetables need more phosphorus (P) and root vegetables need more potassium (K). For soil deficient in nitrogen, phosphorus and potassium (NPK), fertilizers commonly used as a remedy to these deficiencies are: Urea for Nitrogen deficiency; cultivation; Triple Super Phosphate (TSP) for phosphorus deficiency ; Muriate of potash (MP) for potassium deficiency; Sulphur (S) and zinc (Zn) fertilizers are also being used to obtain maximum benefit from NPK fertilizers.2.15.1 There are two basic types of fertilizers: Inorganic or Organic

2.18 Inorganic (Chemical) Fertilizers

These are always manufactured in industries. Chemical fertilizers must be applied in balanced doses. Inorganic fertilizers are not always locally available and tend to be expensive. Various types of fertilizers are:

Simple fertilizers- such as Urea, , Di-ammonium phosphate (DAP), Single super phosphate(SSP), Triple super phosphate (TSP) and Muriate of potash(MP) supply only one nutrient; *Mixed fertilizer* – contains more than one but not all of the nutrients, which the plants need; *Complete fertilizers* – supplies all the basic nutrients to plants.

2.18.1 Nitrogen

Functions: It increases leaf and stem growth and gives dark- green colour to plants. It regulates use of phosphorus and potassium fertilizers; and makes plants less susceptible to attack by pest and diseases. Along with, it increases protein content of some food and feed crops.

Deficiency Symptoms: Leaves become pale -green and then turn yellow. Growth and development of leaves is poor; stalks and branches stunt; and roots, shoots and fruits are smaller.

Nitrogen Fertilizers Include the Following:

i)   **Anhydrous ammonia (NH_3):** It contains 82% N; gas at the normal pressure and liquid under high pressure. Special equipment is needed to inject it into the soil to avoid losses through evaporation (volatilization). High pH increases volatilization and soil- moisture content below and above field capacity also increases loss of this nutrient. Heavy soils retain ammonia better than light soils. The high amount of organic matter also increases ammonia retention. Although the initial effect is to increase *p*H; ammonia ultimately decreases it.

ii) Non-pressure nitrogen solutions: It contains 40-43% liquid N. It is produced from urea and ammonia nitrate solutions plus water. It is commonly referred to as UAN solution. It is easier to handle and apply than other nitrogen materials, such as ammonia or dry products. It can also be applied more uniformly and accurately through various irrigation systems. However, it can cause burning injury when sprayed on leaves.

iii) Urea: It is a popular, dry N carrier. It is more expensive than NH_3 because of the transportation cost and additional production steps involved to increase its nitrogen content (46%); it can be stored, handled, and applied without special equipment. The N in the urea is lost through volatilization. About 20%-30% of the N is lost when urea is applied on the soil surface and is not moved into the soil by water. When added to the soil, urea hydrolyzes to ammonium carbonate, which readily decomposes to NH_3 and carbon dioxide. The hydrolysis is enhanced by enzyme urease found in soil.

iv) Ammonium nitrate: It supplies both ammonium and nitrate ions. Its nitrogen content ranges from 33to 34%.It has excellent handling qualities, readily absorbs moisture, is chemically stable and is a good source of sulphur. The sulphate ions make this fertilizer more acidic than ammonium nitrate, and its continued use without lime would decrease soil *p*H to a level that may reduce soil productivity. However, it can be useful in high-*p*H soils and for acid-requiring crops.

v) Ammonium phosphate: This supplies phosphorus in addition to nitrogen, and both are water-soluble; hence, it is preferred in manufacturing liquid fertilizers.It contains 12-15% N and 25-27% P.

vi) Slow-release nitrogen compounds: Most of the nitrogen in fertilizers is easily lost through volatilization, leaching, and plant absorption. To reduce nitrogen losses, some fertilizer materials, like urea-formaldehyde compounds and sulphur-coated urea compounds, have been developed. These slow-release nitrogen compounds contain 32-38% nitrogen.

2.18.2 Phosphorus

Functions: It makes plants more drought-resistant and hardy, hastens maturity and helps seed and fruit formation.It also increases legume growth, helps legume nodule formation and stimulates root formation.

Deficiency Symptoms: Shoots and roots are stunted. Lateral shoots and buds are poorly formed. Flowering is reduced and bud and leaf opening is delayed. Resistance to diseases decreases. Blush-green colour is visible on the leaf surface.

The main source of phosphorus fertilizers is rock phosphate (the mineral apatite). The rock can be acid or heat-treated to make phosphorus soluble.

i)   Phosphoric Acid: The material used for this fertilizer is produced by treating rock phosphate with sulphuric acid and it contains 54% P_2O_5. It can be injected directly to the soil; but since a special equipment is needed, the acid is more frequently added to irrigation water instead.

ii)   Superphosphate: It is a traditional phosphorus fertilizer produced by treating rock phosphate with small amount of sulphuric acid, and contains 16-20% available P_2O_5. Although its phosphorus content is low, it is high in sulphur and calcium.

iii) Ammonium Phosphate: It is produced by reaction of ammonia with phosphoric acid or by combining phosphoric and sulphuric acids. Di-ammonium phosphate (18% N and 46% P_2O_5) is used in large quantities especially in the manufacturing of high analysis fertilizers. Ammonium phosphate-sulphate (16-20-0) which is completely water-soluble is also widely used. This makes the soil acidic because of ammonia.

iv)   Basic Slag: It is a by-product of steel production, commonly used in European countries and contains readily available phosphorus and is alkaline. It is effective on acid soils, apparently because of its high calcium hydroxide content.

v)   High Analysis Phosphate: Two promising very high-analysis phosphate fertilizers are calcium met phosphate and super phosphoric acid. They are very important materials in the production of liquid fertilizers and other high analysis fertilizers.

2.18.3 Potassium

Functions: It increases resistance of plants to diseases, and produces stiff stalks and stems thus reducing water logging. It helps move food from leaves to root, creates winter hardiness and growth of fruit and root vegetables.

Deficiency symptoms: They are curling of leaves and upper surface of leaf becomes wrinkled. Gradually slight scorching occurs, starting with older leaves. Plant growth is reduced and stunted, with short internodes and bending of main stem towards ground. Fruit often ripens unevenly.

Following are different potassium fertilizers.

i) Potassium chloride: It is a commercially known as muriate of potash(common name of hydrochloric acid). It contains 40-50% potassium (48-60% K_2O) and is most widely used potassium fertilizer. It is used in the soil (direct application) and for manufacturinng NPK (complete) fertilizii*ii*. Potassium sulphate: It contains 40-42%

potassium and 48-50% K_2O. It has advantage of supplying sulphur, which is more widely deficient in soils than chlorine.

ii) Potassium magnesium sulphate: It is a double salt of potassium chloride and potassium sulphate with small amounts of chlorine. 19-25% K. Although rather low in potash, and is useful in soil with magnesium deficiency.

2.18.4 Mixed Fertilizers

It contain at least two of the fertilizer elements, and usually all the three. When the three major elements are present, the fertilizer is called a complete fertilizer. The amounts of the three major elements, N, P, and K, in the fertilizers are indicated in percentage by three numerals designating fertilizer grade. For instance, a 14-14-14 complete fertilizer contains 14%N, 14% P_2O_5 and 14% K_2O. Common grades of complete fertilizer are 14-14-14, 12-24-12 and 5-10-16.

2.18.5 Sources of Micronutrients

- **Boron:** The most common source of boron is sodium tetra borate. Some boron fertilizers are: Fertilizer Borate-68 (21% B) and Solubor (20-21% B).
- **Cobalt:** Cobalt sulphate is a source of this element. A cobaltized super phosphate is also used.
- **Copper:** The usual source of copper is the copper sulphate salt which contains 25.5% sulphur. Copper ammonium phosphate (30% copper) is used for soil application or as a foliar spray.
- **Iron:** Ferrous sulphate with 19% iron is commonly used as a spray solution in foliar applications. Iron chelates (5%-14%) also are widely used.
- **Manganese:** Manganese sulphate (26%-28% Mn) is commonly used to correct deficiencies of the element. It is applied on the soil or on leaves.
- **Molybdenum:** Ammonium molybdate (54% Mo) and sodium molybdate (30% Mo) are the sources of this element. Molybdenum trioxide (66% Mo) is incorporated in the pelleting of seeds.
- **Zinc:** Common material is zinc sulphate with about 35% zinc. Chelates of zinc are also available sources with 9%-14% zinc.
- **Chlorine:** The chlorides of ammonium, calcium, magnesium, potassium, and sodium contain readily available chlorine, which ranges from 49% to 74%.

Advantages: Nutrients in them are soluble and quickly available.Formulations can contain relatively high amounts of total nutrients. Small amounts can be applied to provide needed nutrients. Inorganic fertilizers are relatively inexpensive.

Disadvantages: Over fertilization may occur, since often only small amounts are needed. Soluble nutrients in concentrated solutions may be caustic to growing plants. Some nutrients are very soluble and may be lost from the plant root zone through leaching.

2.19 Organic Fertilizers

They are used to maintain fertility and increase production. The most common sources of organic fertilizers are animal manure and urine, crop by-products and the remains of dead plants. Organic fertilizers contain all the nutrients needed by plants. It is inexpensive and can be prepared or procured at any time. The materials required for organic fertilizers are always available locally.

Advantages: Increases soil fertility, supplies nitrogen, phosphorus, potassium and other nutrients. Increases water-holding capacity of the soil which is important for sandy or coarse loamy soil, where rainfall is marginal. Increases population and activity of soil micro-organisms, which help improve nitrogen fertility status. Improves cation exchange capacity (CEC) of the soil, which becomes the storehouse of plant nutrients. Prevents compactions, reduces soil erosion and supplied more oxygen to roots. Improves physical conditions of the soil, making it easier to cultivate. Helps in effective use of chemical fertilizers. Reduces toxicity of the soil caused by chemical fertilizers and insecticides.

Various Types of Organic Fertilizers are

2.19.1 Farmyard Manure (FYM)

It is produced by decomposing together or individually dung and urine of domestic animals, poultry litter, straw, ash and sawdust.

2.19.2 Dung Fertilizer

It is produced by preserving animal urine and dung. Dung contains NPK and organic matter. It is generally available to all farmers. The preservation of dung is important as exposure to direct sunlight or rainwater reduces its quality.

2.19.3 Compost

It is produced from such materials as rotted house waste (including kitchen scraps), waste straw, dead plants, water hyacinth, leaves, weeds, plant stalks, fruit skin, eggshells, crop residues, animal manure and urine, poultry litter and fish waste. Vermi compost is used to properly decompose agricultural waste and FYM with the help of earthworm species under the moist conditions.

2.19.4 Cover Crops/Green Manure

It is produced by decomposition of cultivated crops and ploughed soil. There are many green manuring and leguminous crops which supply plant nutrients (especially nitrogen) to soil such as pulses and beans.

A legume-grass mixture is an effective green manure crop. The thick growth of the cover crop also helps smoother weeds.

2.19.5 Peat Moss

It is generally acidic and is useful in all soils as an amendment for acid-tolerant crops. The most effective types are formed from either sphagnum moss or reed sedges. These decay slowly in the soil and hold a considerable amount of water. Peat-humus materials decay completely and tend to decompose more rapidly than peat moss, once incorporated into the soil. Peat moss is useful for landscape plantings because of its longevity and effectiveness in the soil. However, it costs more than most other forms of organic materials, which limits its use to small areas.

Other Organic fertilizers include: bone meal, fish meal, blood dust, ash, rotted leaves and cakes made from mustard, peanuts, coconut or sesame/til.

Nitrogen, Phosphorus and Potassium

Nitrogen, phosphorus and potassium content of different organic materials commonly available in rural areas households and with their nutrient content are as follows.

Table 2.1: Percentage of NPK in Different Types of Manures

Manure	Nitrogen (N)	Phosphorus (P)	Potassium (K)
Cow (old dung)	2.41	0.75	0.88
Buffalo	1.10	0.82	0.70
Broiler	3.17	3.29	2.41
Layer	2.85	4.21	2.00
Goat	1.50	1.50	2.00

Table 2.2: Percentage of NPK in Different Organic Materials

Organic Material (%)	Nitrogen (N)	Phosphorus (P)	Potassium (K)
Water Hyacinth Compost	3.00	2.00	3.00
Compost (general)	0.60	0.48	0.85
Town Compost	1.40	1.00	1.40
Sun Hemp	0.75	0.12	0.51
Dhaincha	0.62	--	--
Cowpea	0.71	0.15	0.58
Mustard Cake	5.20	1.80	1.20
Sesame/Til Cake	6.20	2.00	1.20
Tishi Cake	5.50	1.40	1.30
Groundnut/Peanut	7.10	1.50	1.50
Coconut Cake	3.00	1.90	1.50
Neem Cake	5.30	1.10	1.40
Leaves (rotten)	0.75	0.50	1.26
Bone Meal	3.00	28.00	--
Fish Meal	7.00	6.00	1.00
Blood Dust	11.00	1.50	0.60
Ash	--	2.00	5.00
Rice Straw	0.52	0.52	1.61
Wheat Straw	0.63	0.46	0.86

Table 2.3: Assessment of Available Materials as Source of NPK

Material	Nitrogen (N)	Phosphorus (P)	Potassium (K)
Rice Bran	2	1	1
Wheat Bran	2	1	1
Peanut Shells	1	1	1
Eggshells	1	1	1
Sugar by-product	1	3	1

Contd...

Material	Nitrogen (N)	Phosphorus (P)	Potassium (K)
Feathers	3	1	1
Tea Leaves	2	1	1
Banana Stalks	1	2	3
Banana Leaves	1	2	3
Banana Skins	1	2	3
Corn Cobs	1	1	3
Corn silage	1	1	1
Corn Stalks	1	1	1

1 = Poor Source, 2 = Fair Source, 3 = Good Source

2.20 Improving Soil with Organic Matter

Organic matter, or humus, is a valuable part of soil. It is the end product of decaying organic matter and the most effective material for improving tilth. When incorporated into soil, humus produces a spongy texture that increases soil water-holding capacity. It provides needed pore space, which lets in the air essential to good plant growth. It prevent tiny particles of clay from cementing themselves into a solid mass when wet or dry, thus making soil more easily penetrable by plant roots. It fills in excess pore space in sandy soil, slowing its drainage rate and increasing water-holding capacity. It regulates soils temperature. It releases small amounts of nitrogen and other nutrients for plant use through decay. It increases cation exchange capacity (a measurement of soil's ability to hold nutrients), so that soils can hold and release more nutrients. It also promotes growth of microorganisms, which help condition the soil.

Organic matter may be added to soils in the form of manure, compost, peat moss, peat-humus, spent mushroom compost, and composted sawdust. Very coarse forms of organic matter such as chopped brush or shredded tree bark, should be composted before incorporation.

Alternatively, organic matter can be produced in a vegetable garden by planting winter cover crops, green manure crops, when the land is not being used for gardening. This is an effective way to improve soil condition.

However, organic materials used alone seldom supply a balanced source of nutrients. Most are low in phosphorus, and decaying straw, leaves, grass clippings, and sawdust can temporarily deplete soil of available nitrogen. Reduced amounts of available nitrogen can damage some short-season vegetables.

Add no more than 5 cm of organic matter to garden soil each year. Use much less if the organic matter is rich in nutrients such as fresh animal manure. Organic matter of soils contains organic residues and living organisms. Adding un-decayed organic residues stimulates rapid increase in microbial activity in the soil. This can temporarily tie up the available nitrogen (N) supply until soil organisms have attained a new equilibrium.

Advantage: They are less costly than inorganic fertilizers when used in large application. Nutrients are available more slowly over a longer period of time. Nutrients are less likely to be leached from the soil. Organic fertilizers may act as soil amendment and conditioner.

Disadvantage: Organic fertilizers generally cost more than inorganic. Many organic fertilizers are low in nitrogen and other plant nutrients. Some forms need to be composted before use. Nutrients in them are in insoluble form before soil microorganisms break them down.

2.21 Factors Determining Fertilizer Needs

2.21.1 Crop Type

Economic value, nutrient removal, and absorbing ability of the crop should be considered. High-value vegetables can be fertilized heavily to ensure high yields.

Excess nutrients applied are not wasted but used to build- up soil fertility of the succeeding crops. A high-nutrient removal needs high fertilizer application to compensate for nutrient loss. The nutrient-supplying power of the soil (fertility) should also be considered in determining kind and amount of fertilizer to be used.

2.21.2 Chemical Condition of the Soil

This is evaluated in relation to total and available nutrients in the soil. The amount of fertilizer needed is difference between the crop's nutrient requirement and the amount supplied by the soil. It is difficult to quantify, since plant and soil are constantly changing and interacting with many other factors. Field experimentation, soil testing, plant analysis, and information on deficiency symptoms can be used as a basis for reliable fertilizer recommendation for a specific crop and soil type.

Time of application: The kind of crop, climate, soil and nutrient can influence time, frequency and amount of fertilizer application. Vegetable crops differ in their growth and development patterns. Moreover, each crop has a different growth duration. The need for nutrients varies depending on the crop, especially as regards to growth and development of the plant part of economic importance. **Rainfall** affects availability of nutrients to plant; between the time it is applied and the time the nutrients are used by the plant. **Temperature** also affects release of nitrogen, phosphorusand sulphur from organic matter. Likewise, it affects nitrification and absorption of phosphorus and potassium by plants. **Soil factor** is important because soils have different percolation rate, fixing capacity and nutrient availability. Fertilizers are applied at planting or before planting (basal application). Fertilizers from organic sources are applied much earlier; so that they decompose partially, and the nutrients are available to plants. Fertilizers are applied not only at planting but also during growing season (side-dressing or top dressing).

2.22 Method of Application

Fertilizers should be placed in the soil zone where it serve plant to its best advantage. There are three important considerations in determining the proper application method—efficient use of nutrients from the time of plant emergence to maturity; prevention of salt injury to seedlings and convenience of grower. The method of application, therefore, should see to it that nutrients are available to plants at all times during its growth. The right amount of nutrients should also be made available to the developing crops. The common methods of application are as follows.

- **Broadcast**: The fertilizer is applied uniformly over the plots before planting. It is then incorporated by tilling or cultivating.

- **Banding**: The fertilizer is applied in bands on one side, both sides, or below the seeds or transplant. Care should be taken not to injure seedlings through contact with fertilizers.

- **Topdressing or side-dressing**: Topdressing means broadcasting fertilizers on the crop, while side-dressing means applying them besides crop rows. Both are done after crop emergence.

- **Fertigation**: This is the application of fertilizers through irrigation water. Nitrogen and sulphur are principal nutrients commonly used. Potassium and highly soluble forms of zinc and iron can also be readily applied this way. Phosphorus and anhydrous ammonia may precipitate in water with high calcium and magnesium contents so they are not used in fertigation.

- **Foliar application**: This method can be used with fertilizer nutrients readily soluble in water. It is also used when there is a soil-fixation problem. In this method, however, it is difficult to apply sufficient amount of major elements. Nutrient concentrations of 1 to 2% can be applied without injury to foliage. This method, therefore, is commonly used only to apply minor elements or supplements of the major elements.

Drenching: Keeping in view that farmers urgently need judicious fertilization strategies which can improve efficacy of nutrient uptake by plants and minimize environmental risks, this AVRDC's developed starter solution techniques are helpful in enhancing early fruit- setting, yields as well as quality in tomato, capsicum, cucumber, pepper, eggplant, cabbage and leafy vegetables etc. However, the effect may vary with respect to potting media for nursery raising and soil property of the location. Water soluble fertilizer (WSF) is a mixed fertilizer (20%N: 20%P: 20%K) and used in field as a source of multi-nutrients. The starter solution is prepared by calculating quantity of WSF in kg/ha and applied at the time of transplanting as well as at four various critical stages of growth.

Getting most from fertilizers: To make the best use of fertilizers, test soil every three to five years, select crops suitable for the soil, be sure soil is well drained, apply lime and fertilizers as indicated by soil testing, control weed growth, use adequate organic matter to improve soil quality, avoid overcrowding of plants in vegetable gardens and use disease-resistant varieties of seeds and plants.

2.23 Composting for Home Gardens

2.23.1 What is Compost?

Compost is decomposed organic material which is made with materials such as leaves, shredded twigs, and kitchen scraps from plants. Compost is considered "black gold" because of its many types of benefits. Adding compost to clay soils makes them easier to work and plant. In sandy soils, addition of compost improves water- holding capacity of the soil. Compost can help improve plant growth and health. It is also a good way to recycle leaves and other yard waste.

2.23.2 Composting Process

Primary variables that must be "controlled" during composting include:

- **Particle size:** Grinding, chipping and shredding materials increase surface area on which microorganisms feed. Smaller particles produce a more homogeneous compost mixture and improve pile insulation to help maintain optimum temperature. If the particles are too small, they may stop free flow of air through the pile.

- **Moisture content:** Microorganisms living in a compost pile need adequate amount of moisture to survive. Water is the key element that helps transport substances within the compost pile and makes nutrients in organic material accessible to microbes. Organic material contains some moisture, but moisture also might come in the form of rainfall or intentional watering.

- **Oxygen flow:** Turning the pile, placing the pile on a series of pipes, or including bulking agents such as wood chips and shredded newspaper all help aerate the pile. Aerating the pile allows decomposition to occur at a faster rate than anaerobic conditions. Too much oxygen can dry out pile and impede composting process.

- **Temperature:** Microorganisms require a certain temperature range for optimal activity. Certain temperatures promote rapid composting and destroy pathogens and weed seeds. Microbial activity can raise the temperature of the pile's core to at least 60 °C. If the temperature does not increase, anaerobic conditions (i.e., rotting) occur. Controlling earlier four factors can bring about proper temperature. Frequent turning helps speed composting. Once the pile has cooled in the center, decomposition of the material has taken place.

- **Surface area:** It can be maximized by shredding and chipping all clippings and waste into small pieces with a chipper/shredder. The more surface area you expose for microorganisms to attack, the faster is the decomposition.

- **Carbon to nitrogen ratio:** Organic materials rich in nitrogen are referred to as GREENS (fresh veggie scraps or grass clippings), while the others can be lumped together as BROWNS (hay, twigs, dried leaves) ,which are rich in carbon. A good rule of thumb is to use 2-3 parts brown to 1 part green. Alternating layers ensure proper mixing. For piles mostly having brown material (dead leaves), try adding a handful of commercial 10-10-10 fertilizer to supply nitrogen and speed-up composting.

2.23.3 Time Taken for Composting

The amount of time required to produce compost depends on size of the compost pile, types of materials, surface area of the materials, and number of times the pile is turned. For most efficient composting, the pile should be between 90 cm cubed and 150 cm cubed. This allows the centre of the pile to heat up sufficiently to break-down materials. Smaller piles can be made but will take longer to produce finished compost. Larger piles can be made by increasing length of the pile but limiting the height and the depth to 5 feet tall by 5 feet deep. If the pile has more brown organic materials, it may take longer to compost. Composting can be sped up by adding more green materials or fertilizer with nitrogen (use one cup per 25 square feet). By breaking materials down into smaller parts (chipping, shredding, mulching leaves), the surface area of the materials would ncrease. This helps bacteria to break down materials more quickly into compost.

By turning frequently (about every 2-4 weeks), more compost can be produced quickly. Waiting at least two weeks allows the centre of the pile to heat -up and promotes maximum bacterial activity. The average composter turns pile every 4-5 weeks. With frequent turning, compost can be ready in about 3 months, depending on the time of the year. In winter, activity of the bacteria slows down, and it is recommended to stop turning pile after November to keep heat from escaping from pile's centre. In summer, warm temperatures encourage bacterial activity and composting is quicker.

2.23.4 Layering Compost

Layering is the recommended method for starting a compost pile. Layering is adding thin, uniform layers of materials in a repeated pattern. Start the compost pile on bare ground, removing sod or existing vegetation. Contact with the soil would provide bacteria needed for composting. Do not place the pile on concrete.

Layer 1: Add a 15-20 inch layer of organic matter, both brown and green. Do not pack materials in, as this limits air flow and oxygen needed by bacteria.

Layer 2: Add a starter material, such as animal manure, fertilizers or commercial starters. These materials help to heat up pile by providing nitrogen for bacteria and other microorganisms. Select one of the following: 2.5-5 cm layer of fresh manure from a grain- eating animal, or 1 cup of 10-10-10 or 12-12-12 fertilizer per 25 square feet, or a commercial starter (follow label directions).

Layer 3: Add a 2.5-5 cm layer of top soil or finished garden compost. This is done to introduce microorganisms to the pile. Avoid using soil recently treated with insecticides and also avoid using sterile potting soil.

2.23.5 Materials to Compost

What to use: Leaves, some manures (cow, horse, sheep, poultry, rabbit), lawn clippings, vegetable or fruit wastes, coffee ground, shredded newspaper or white, unglazed office paper, trimmed plant materials, shredded stems and twigs, cardboard rolls, clean paper, cotton rags, dryer and vacuum cleaner lint, eggshells, fireplace ashes, hair and fur, hay and straw, houseplants, nut shells, wood chips and wool rags etc.

What not to use: Glazed, colour printed magazine paper, black walnut tree leaves or twigs , released substances ,which might be harmful to plants, coal or charcoal ash may alsot contain substances harmful to plants, dairy products (e.g., butter, egg yolks, milk, sour cream, yogurt), other plants, fats, grease, lard, or oils create odour problems and attract pests such as rodents and flies, meat or fish bones and scraps, pet wastes (e.g. dog or cat feces, soiled cat litter), may contain parasites, bacteria, germs, pathogens, and viruses harmful to humans, yard trimmings treated with chemical pesticides, may kill beneficial composting organisms.

2.24 Methods of Composting

2.24.1 Backyard or on Site Composting

Backyard or on site composting is suitable for converting yard trimmings and food scraps into compost that can be applied on the site. Backyard or on site composters also might keep leaves in piles for eventual use as mulch to retain moisture. Climate and seasonal variations do not present major challenge to backyard or on site composting because this method typically involves small quantities of organic wastes. Improper management of food scraps can cause odours and also might attract unwanted attention from insects or animals. The conversion of organic material to compost can take up two years, but manual turning can hasten the process considerably (e.g. 3 to 6 months). The resulting natural fertilizer can be applied to home gardens to help condition soil and replenish nutrients.

2.24.2 Vermicomposting

It uses worms to convert organic waste to compost rather than microbial-dependent decomposition process used in backyard composters. Vermicomposting enriches soil, improves its water retention, and enhances germination and plant growth. Through this method, red worms are placed in bins with organic matter to break it down into high-value compost, called castings. Worm castings are loaded with nutrients as they contain seven times more phosphorus, five times more nitrogen, and eleven times more potassium than typical soil. These help retain moisture in the soil and enhance growth and yield of the garden. Worm bins are easy to construct (they are also commercially available) and can be adapted to accommodate volume of food scraps generated. Worms would eat almost anything in a typical compost pile (e.g., food scraps, paper, plants etc.). Worm composting can be done indoors, which allows for year-round composting, as well as composting in small places like apartments, classrooms, or homes without large yards. It is frequently used in schools to teach children conservation and recycling. Another winning benefit: as worms move through the bedding and compost, they aerate pile, eliminating need for manual pile turning.

Worms are sensitive to variations in climate. Extreme temperatures and direct sunlight are not healthy for them. The optimal temperature for vermicomposting ranges from 13 to 25 °C. In hot, arid areas, the bin should be placed under the shade. Worms grow well in a carbon to nitrogen ratio of about 30:1, plus water to keep it moist. The primary responsibility is to keep the worms alive and healthy by providing proper conditions and sufficient food. There are two methods of feeding: **top feeding** and **pocket feeding**. Top feeding means organic materials are placed directly on top of the existing layer of bedding. Pocket feeding is when a top layer of bedding is maintained and food is buried beneath.

Requirements for vermicomposting include: worms, worm bedding (e.g., shredded newspaper, cardboard), and a bin to contain the worms and organic matter. Maintenance procedures include preparing bedding, burying garbage, and separating worms from their castings. Vermicomposting bins should have holes on the side for aeration so that the bin is ventilated and the worms can breathe. Without enough oxygen, compost will become anaerobic decay and would produce offensive odour. Approximately 800-1,000 mature worms can eat up to 226g of organic material per day. It typically takes three to four months for these worms to produce harvestable castings, which can be used as potting soil. Vermicomposting also produces compost or "worm" tea, a high-quality liquid fertilizer for house- plants or gardens.

Harvesting castings: Dump and Sort method requires pouring compost bin contents onto a plastic sheet or a similar, waterproof platform under bright light. Separate contents into pyramid- shaped piles. The photo-sensitive worms make their way to the bottom of the piles, and in 10-15 minutes, the rich castings can be skimmed from the top of the pyramids. Repeat until only the worms remain, then place them into the bin with fresh bedding to start vermicomposting again.

The **"Side by Side"** method is recommended for squeamish who prefers not to touch worms. Begin the process by burying organic scraps in different spots on one side of the bin over a number of weeks. The worms would migrate to the side with food, and gardeners are now free to cull castings on the other side. When it's ready to harvest again, place food on the opposite side and repeat the process.

Fig. 2.1: Farmer Showing Vermicomposting Pit

2.24.3 Factors Affecting Vermicomposting

i) **Worms:** They can consume in carbon and nitrogen rich organic scraps every 24 hours. A good way to calculate how many worms to buy is to use a 2:1 ratio—two kg of worms for every kg of organic- kitchen scrap.

ii) **Temperature:** The ideal temperature for the bin is between 16 and 27°C, but a wiggle room of 4-32°C before worms' impacts adversely. It is best to store bin in a cool, dark place within the home for stability. Do not allow the bin to freeze or overheat.

iii) Bedding (carbon content): Bedding for worms is important. Initially, it takes up approximately 2/3 of the new bin space and provides half of the worm its carbon-nitrogen diet. In addition, it offers a dark, moist hiding place for photosensitive worms. Suitable bedding materials are shredded cardboard, paper, coir bricks, untreated wood shavings, and chopped straw and hay.

iv) Water content: If worms dry out, they die. For this reason, bedding must be kept moist but never dripping wet, as anaerobic (oxygen-free) conditions can occur and lead to odours and suffocation of worms (they can drown). Since worms themselves produce liquid, it may be necessary to occasionally add dry bedding to the bottom of bins that have become saturated. It is to make sure that the bin has holes on the bottom to allow adequate drainage.

v) Food scraps (nitrogen content): Chop up organic kitchen scraps and add them to worm bin composter to help worms digest their meal. Feed worms twice a week, between 650 and 900g of accumulated scraps, adjusting portions to the amount of worms as required using 2:1 ratio.

vi) Oxygenation: Worms require a constant source of fresh air, breathing through their skin. There is no need to turn compost to aerate.

2.24.4 General Steps of Vermicomposting

- Choose a proper composting bin or a specific site.
- Prepare bedding for worms, making sure it takes up at least 2/3 of the container space and is moist but not soggy.
- Add earthworms to the bedding under direct light. Due to their photosensitivity, the worms will begin to burrow into dark safety of the bedding.
- After the worms have been allowed to settle for a day or two, begin feeding them with organic scrap.
- When bedding has been almost entirely consumed, harvest valuable compost.

2.24.5 What to Put in Worm Bin?

Fruit and veggie scraps, leftover vegetables (without oil or sauce), plain rice, egg cartons, coffee trays, tea bags (without staples), coffee grinds (for grit), breads, grains and cereals, beans, untreated sawdust, grass clippings, hair clippings, plant trimmings ,paper and leaves.

2.24.6 What Not to Use?

Garlic, Onion, meats and bones, dairy products, oily food, heavily spiced or hot foods, anything with insecticide or chemicals, twigs, metal and foils, plastics, weeds and manure etc.

2.24.7 In-vessel Composting

Organic materials are fed into a drum, silo, concrete-lined trench, or similar equipment where environmental conditions—temperature, moisture and aeration—are closely controlled. The apparatus usually has a mechanism to turn or agitate material for proper aeration. In-vessel composting can process large amounts of waste without taking up as much space as the windrow method. In addition, it can accommodate virtually any type of organic waste (e.g., meat, animal manure, biosolids, food scraps). This type of composting can be used year-round in virtually any climate because environment is carefully controlled, often by electronic means. In this composting, very little odour and minimal leachate are produced. In-vessel composters are expensive and might require technical assistance to operate properly, but this method uses much less land and manual labour. Conversion of organic material to compost can take as little as a few weeks. Once the compost comes out of the vessel, it still requires a few more weeks or months for microbial activity to stabilize and the pile to cool.

2.24.8 Where to Place the Compost?

For the compost bin to function properly, place the compost pile in an area with good air circulation. The pile should not be in direct contact with wooden structures, as this will cause decay. It is best to locate the pile in partial shade, but this is not a necessity. It may be located close to the home garden and the water source. If kitchen scraps are added regularly, it may be more convenient to have the pile near kitchen. It can be screened from view with a fence or by placing it behind shrubs or a taller structure.

2.24.9 Types of Compost Bins

i) Holding units: It requires low maintenance, good choice for those with limited space such as apartment dwellers, do not require turning, however the lack of aeration causes composting process to take 6 months to 2 years.

ii) Portable bins: Its similar to holding units, except that they can be taken apart and moved;materials can also be mixed with this type of bin.

iii) Turning units: These are designed to be aerated. They produce compost faster because they supply oxygen to bacteria in the pile. These units may also have less odour problem, which is associated with poor aeration; it may be either a series of bins or a structure that rotate, such as a ball or barrel, often cost more and are more difficult to build. Once these units are filled and turning process begins, new materials should not be added.

iv) Heaps: It is an option for those who do not wish to build or purchase a bin structure. Turning heap is optional, but composting process will be slowed if pile is not turned. Woody materials may take a very long time to decompose with this method, and food scraps may attract pests.

v) Sheet composting: In this composting is done during the fall, when thin layer of materials such as leaves (that have not been composted) are worked into the garden. By spring, the material would be broken down. The decomposition process ties up soil nitrogen, making it unavailable to other plants.

vi) Trench composting: Organic material is buried in holes/trench 20-37.5- cm deep, and then covered with soil dug from the hole. Decomposition takes about a year, as limited oxygen slows the process. It is recommended to avoid planting that area for a year, as the nitrogen available to plants may be limited by decomposition process.

2.24.10 Benefits of Composting

i) Enriches soils: Composting encourages production of beneficial micro-organisms (mainly bacteria and fungi), which in turn break down organic matter to create humus. Humus--a rich nutrient-filled material--increases nutrient content in soils and helps them retain moisture. Compost has also been shown to suppress plant diseases and pests, reduce or eliminate need for chemical fertilizersand promote higher yields.

ii) Clean-up (remediate) contaminated soil: The composting process has been shown to absorb odours and treat semi-volatile and volatile organic compounds (VOCs), including heating fuels, polyaromatic hydrocarbons (PAHs) and explosives. It has also been shown to bind heavy metals and prevent them from migrating to water resources or being absorbed by plants. The compost process degrades and, in some cases, eliminates wood preservatives, pesticides, and both chlorinated and non-chlorinated hydrocarbons in contaminated soils.

iii) Prevent pollution: Composting organic materials which have been diverted from landfills ultimately stop production of methane and leachate formation in landfills. Compost prevents pollutants in storm water runoff from reaching surface water resources. Compost also prevents erosion.

iv) Economic benefits: Using compost can reduce need for water, fertilizers and pesticides. It serves as a marketable commodity and is a low-cost alternative to standard landfill cover and artificial soil amendments.

Compost can also facilitate reforestation, wetlands restoration, and habitat revitalization efforts by amending contaminated, compacted, and marginal soils. Capture and destroy 99.6% of industrial volatile organic chemicals (VOCs) in contaminated air.

v) Composting prevents insect-pests: The uses of composting in crop cultivation not only enhances organic carbon but also enhances beneficial microorganisms and thus reduce insect-pest and disease occurrence in soil and above ground.

Fig. 2.2: Ready Composting Material and Compost

2.24.11 Application

i) Top dressing: It refers to compost being spread around soil during growing season. Perfect for herb and vegetable gardening, compost amends garden soil with nutrients, which allows plants to foster healthy growth while keeping plant diseases and insect problems at bay. Poor soil can be augmented with 2-3 inches of compost; and one inch of compost is thick enough for spreading on garden beds.

ii) Side dressing: It is a good option when running low on compost. Spread compost on certain vegetables or rows. Work the compost into the soil around the vegetable, starting about 2.5 -cm from the stem.

iii) Compost as mulch: Compost can also be used as mulch, which is meant to cover all of the soil around the plant. Mulches protect from soil erosion and help retain water while adding nutrients. Spread about 1.5-2.5- cm thick layer of compost on bare soil and then cover with a 5-7.5 layer of mulch to nourish vegetables.

2.25 Climate

The growing season is essentially the length of the time an area can give plants conditions they need to reach maturity and yield produce. It is totally dependent on the local climate. The way a vegetable type reacts to climatic conditions — heat, cold, moisture, and so on — determines its "hardiness". The vegetables that are grown in a home vegetable garden fall into one of four hardiness categories: *very hardy, hardy, tender* and *very tender.*

2.25.1 Temperatures

Average day-to-day temperature plays an important role in the growth of vegetables. Temperature, both high and low, affects growth, flowering, pollination, and development of fruit. If the temperature is too high or too low, leafy crops may be forced to flower prematurely without producing desired edible foliage. This early flowering called "going to seed," affects crops like cabbages and lettuce. If the night temperatures get too cool it may cause fruiting crops to drop their flowers — reducing yields considerably; peppers may react this way to cold weather. Generally, the ideal temperature for vegetable plant growth is between 40° and 85°F. At warmer temperatures, plant's growth would increase, but this growth may not be structurally sound. At lower temperatures, plant's growth will slow down or stop altogether. Vegetables have different temperature preferences and tolerances and are usually classified as either cool-season crops or warm-season crops.

2.25.2 Rainfall

The amount and timing of rainfall in the area also affects how the vegetables would grow. Too much rain at one time can wash away seeds or young seedlings and damage or even kill mature plants. And constant rains when certain plants are flowering can reduce pollination of flowers and reduce yield. This can happen to tomato, pepper, beans, eggplant, melons, pumpkin, and both summer and winter squashes. Consistent rains can also force honeybees to stay in hives instead of pollinating plants; affecting yields. Too little rain over a period of time can slow down plant growth and kill young seedlings or even mature plants. Limited moisture in the air can also inhibit pollination and reduce yields of some vegetables.

2.25.3 Sunlight

It provides energy that turns water and carbon dioxide into sugar that plants use for food. Green plants use sugar to form new cells, to thicken existing cell walls, and to develop flowers and fruits. The more intense the light is the more effective it is. Light intensity, undiminished by obstructions, is greater in the summer than in winter, and greater in areas where the days are sunny and bright than in areas where its cloudy, hazy or foggy. As a rule, the greater is the light intensity the greater is plants sugar production provided, of course, that it's not too hot or too cold; the plants must get the right amount of water. If a plant is going to produce flowers and fruits, it must have a store of energy beyond what it needs just to grow stems and leaves. If the light is limited, even a plant that looks green and healthy may never produce flowers or fruits e.g. tomato. All vegetables need a certain amount of light to grow properly, without which they would not flourish. Most vegetables need full sun for best growth, but young or newly transplanted plants may need some protection from bright, direct sunlight. To provide shade where there's too much sun, plant large, sturdy plants to provide a screen, and the garden can be so designed that large plants and small ones get the light they need. Young plants can also be shaded with boxes or screens when necessary.

2.25.4 Day Length

Many plants, including tomato, are not affected by day length (how long light stays during the day). But for many others , length of the day plays a big part in regulating when they mature and flower. Some plants are long-day plants, which means they need 12 or more hours of sunlight daily to initiate flowering. Radish and spinach are long-day plants. These crops go to seed very fast in the middle of the summer when the day length is more than 12 hours. Onions are long-day plants; they stop making leafy growth and start developing into a bulb when day length is 16 hours or more. The more leafy growth that the plant has at this point; bigger is the bulbs. This means that the onions should be allowed to make as much leafy growth as possible before the day length reaches 16 hours. This may be achieved by planting them as early in the season as possible. Other plants are short-day plants and need less than 12 hours of light to initiate flowering; examples are soybean and corn. Some cultivars such as large seed butter (lima) beans are day length neutral and are adapted to long or short days.

2.25.5 Winds

Dry, windy days and cool night temperatures (a 10°F drop from day temperatures) can cause fruiting crops like pepper to drop their flowers before they're pollinated. This can be avoided to some extent by putting up some type of windbreaks to protect crops from drying winds.

Cold frames and hot frames: These can sometimes be used for extending gardening season. A cold frame(a glass enclosed growing area outside) uses solar heat, therefore, it qualifies as an energy-saving device and is often called a "poor man's greenhouse". It's an ideal place to start hardy annuals and perennials or to put plants in the spring to harden them to bear for rigours of outdoor life. When vegetables are grown inside, especially the cold-tolerant ones, one can move them to a cold frame and give them the benefit of much more light in a protected place. On the days when the sun is bright, provide them some shade to save plants from sun burning. One of the advantages of using a cold frame is that you raise a large variety of crops from seed through maturity within the frame, making it possible to have vegetable crops ahead of their normal season when they are extremely expensive to buy. An added bonus is that the cold frame acts as a barrier against insect -pests, birds and small animals. Heat is provided either by rotting manure (the classic system) or by electricity (the modern way). Electricity is much easier but a lot more expensive than manure.

2.25.6 Water

Vegetables are generally more than 90% water. Water is vital from the moment seeds are sown through sprouting to the end of the growing season. Plants need water for cell division, cell enlargement, and even for holding themselves up. Thus, water determines the weight and yield of vegetables. The quality of vegetable products is also determined by the quality of water management. Many defects of vegetable products may be traced directly or indirectly to mismanagement of water supply in production field. Unlike field crops which can be grown under rainfed conditions, vegetables with few exceptions are always irrigated, at least partially. Utmost concern is to use irrigation water in the most efficient way. It is equally important to provide adequate drainage facilities in the field because most vegetables cannot tolerate prolonged waterlogging. In humid tropics, vegetable crops may be classified according to adaptation to wet or dry seasons, roughly corresponding to their adaptation to excess or deficiency of moisture. Dry season, taking all environmental factors into consideration, is generally more favourable for growing vegetables than wet season. Hence all tropically adapted vegetables can be grown successfully during this season, provided irrigation water is available. Without irrigation, lesser vegetables can be grown. Rain- fed dry-season crops are normally limited to those that are early maturing and relatively tolerant to excess moisture during the early stage and drought at later stages. These crops must be sown towards the end of the wet season; so that, enough residual moisture is available for germination and crop establishment. Suitable crops for this type of culture are mungbean, cowpea, radish and early-maturing determinate tomato. With adequate drainage, some crops perform even better during wet season than during dry season. These include yard-long bean, winged bean, and leafy vegetables.

Consequences of water deficiency: There is poor stand when water stress occurs during germination, and yield reduction or decline occurs in quality of fruits such as deformity in of cucumber fruits and beans. Also is calcium deficiency, which causes blossom end- rot in tomato or tip -burn in Chinese cabbage.

Consequences of water excess: It causes leaching of fertilizers, reduced root development and adventitious root development. Also root- rot and other diseases occur, which are favoured by high soil moisture.

Consequences of abrupt changes in water supply: Defects, such as cracking of tomato and bitter- gourd fruits, or carrot and radish roots can be traced directly to fluctuations in moisture supply, which takes place during fruit or root enlargement stage.

2.25.7 Factors Determining How Often to Water a Garden

More water evaporates when temperature is high than when it's low. Plants can rot if they get too much water in cool weather. More water evaporates when the relative humidity is low.

- Plants need more water when the days are bright. Wind and air movement would increase loss of water to atmosphere.

- Water needs vary with the type and maturity of the plant. Some vegetables are tolerant to low soil moisture.

- Sometimes water is not what a wilting plant needs. When plants are growing fast, the leaves sometimes get ahead of roots'. If the day is hot and the plants wilt in the afternoon, one should not worry about them; they will regain their balance overnight. But if plants are wilting early in the morning, water is needed immediately.

Proper water management means applying adequate quantities of water at the right time. It includes: **i) Irrigation** (the delivery of water); **ii) Drainage** (removal of excess water)

a) Determining Irrigation-Water Requirement

Irrigation-water requirement of a crop is determined by consumptive use and irrigation efficiency. Consumptive use or evapo-transpiration is sum of transpiration (water entering plant roots and used to build plant tissues and water being passed through the leaves into the atmosphere) and evaporation water lost to the atmosphere in gaseous form from soil, water and plant surface. Water deposited by dew, rainfall, or sprinkle irrigation and subsequently evaporating without entering plant system is part of the consumptive use. When the consumptive use of the crop is known, the water use of large units can be calculated. Consumptive use is influenced by weather, irrigation practices, length of the growing season, stage of crop development, and other plant factors.

b) Scheduling Irrigation

The frequency and depth of water application are determined by weather and soil conditions, development stage, and depth of root zone specific for crop (or variety) in cool, damp weather and in well-established fields with heavy to loam soil by infrequent (weekly) watering and in newly seeded or transplanted crop in hot, dry conditions in sandy soils by daily watering. Water- holding capacity of the soil determines frequency of irrigation. Sandy soil (coarse particles) holds less water than clayey soils (fine particles); therefore sandy soil must be watered more frequently. Loam soil, a combination of sand, silt, and clay, absorbs water readily and stores it for easy plant use.

c) Rooting Depth of Crops

It determines the amount or depth of irrigation. In an average loam soil, 1 cm of water applied at the surface will wet the soil to a depth of 4-5 cm. The depth will be more in sandy soils and less in clay soils.

Table 2.3: Classification of Crops According to Depth of Root Zone

Crop	Root-Zone Depth (cm)
Shallow-rooted Crops	**Less than 50 cm**
Bush beans	45
Lettuce	30
Onion	30
Radish	45
Medium-rooted crops	**51-100 cm**
Cabbage	60
Carrots	90
Cauliflower	60
Celery	60
Cucumber	90
Eggplant	90
Garlic	60
Pepper	90
Squash	90
Sweet corn	90
Deep-rooted crops	**More than 100 cm**
Melon	155
Okra	110
Snap beans	110
Tomato	120

Close to 70% of soil moisture extracted by roots comes from top 50% of the root zone depth; the portion where root hairs and small roots of the root system are concentrated. For cabbage to have a root zone depth of 60 cm, the soil to be wet by irrigation must be at least 30 cm deep (50% of 60 cm). It may not be necessary to wet the entire root zone depth of 60 cm.

2.25.8 . Methods of Applying Irrigation Water

- Choosing the most suitable method of applying irrigation water depends on: soil texture, topography, water supply and crop.

- Methods: overhead, surface, dripand sub irrigation.

i) Overhead irrigation: Water is applied in the form of spray or artificial rain; mainly by using watering cans.

The water source is usually a shallow well, or a tap- water. It is a labour-intensive method. The nozzle of the watering can consists of a perforated tip. The size of the perforation is small when watering is done on seedbeds and bigger as plant grows. Nozzles are easily detached from the can to facilitate cleaning.

ii) Sprinkler irrigation: It can also be applied through a pipe system under pressure. In the flat lands, regardless of the water source, pressure is generated by pumps. Artificial rain is generated by special devices such as perforated sprinkler lines, rotating sprinklers, or micro-sprinklers. Micro-sprinklers are better suited for growing seedlings than for field production of vegetables. Rotating sprinklers are most common among artificial rain devices for vegetable crops because they are most flexible. It consists of a head with one or more nozzles which is rotated by the actions of the water passing through and which waters a circular portion of the garden around the sprinkler. It is capable of applying water at a relatively slow rate while using relatively large nozzles. Application rates of less than 2.5 mm/hour are possible with these sprinklers. This slow rate is desirable in heavy soils with low infiltration rates (less than 4 mm/hour) and advantageous to small farmer who may be irrigating along with other field activities.

Uses: For crops that require light and frequent irrigations such as bulb crops and leafy vegetables;

- for very closely spaced crops or those which have been replanted by the broadcast method,

- for frost protection,

- for control of some insects, such as thrips, which do not thrive very well on wet foliage,

- for application of fertilizers, pesticides and soil amendment, for temperature reduction during hot days to improve quality and yield of some crops and is best used with cool-season vegetables.

Disadvantage: Initial cost of equipment is high;operating costs are higher than surface irrigation. Maintenance of sprinklers can be a problem if water is laden with debris or silt. In the humid tropics, it may favour development of diseases and weed growth, as it wets the whole field. It reduces efficiency of pesticides applied to foliage. Large evaporation losses are sustained because sprinklers wet entire soil surface as well as leaves of the plants. Winds disturb flow pattern, resulting in unequal distribution of water. It creates humidity in the crop canopy, which leads to foliage disease incidence and severity.

iii) Surface irrigation: It is used for vegetable crops, which do not specifically require frequent and light irrigations such as solanaceous vegetables, cucurbits and legumes. It can be applied to vegetable crops raised on furrow and flooding methods.

Furrow irrigation: It is done by running water through small channels (furrows/trenches) while it moves down or across the slope of the field. The water sips into the bottom and sides of the furrows to provide desired wetting. It is applicable only for row crops in fields with uniform slopes of from 0.25% to preferably not greater than 2.5%. Spacing of furrows is determined by the spacing of the plant rows: one furrow is provided for every plant row. Soil characteristics must also be considered because the lateral movement of water from furrows depends primarily on the texture of the soil; wetting pattern being broader in clays than in sands. **Sandy soils:** for complete wetting, furrows should not be placed more than 50 -cm apart. **Uniform clay soils:** furrow spacing of 120 cm or more. A variation of the furrow method is the use of small rills or corrugation for irrigating closely spaced crops. The space between corrugations is closer in light soils and further apart in heavy soils. The corrugation method can be used on comparatively steep slopes of up to 8% and on irregular fields of uneven topography. It is used primarily on fine-textured soils with low water intake rate and on crops with close spacing.

iv) Flooding method: It is applicable in areas which have flat to uniform and gentle slopes, with abundant and inexpensive irrigation water. The flooding method does not require elaborate land preparation; it is most applicable in fields that have been previously used for paddy rice and for closely-spaced vegetables such as garlic with zero tillage. For vegetable crops, the border strip flooding technique is most applicable. It is one way of controlling flood irrigation to achieve better water distribution and economy. The field is divided into a series of strips, 5-15 m wide and 75-300 m long, depending on the slope and soil texture. It is used only in fields with smooth uniform slopes, preferably not over 3%. It is not desirable on fine-textured soils with low water–intake rate. Another form is moat irrigation which is used in hill crops. A moat surrounds each plant and delivers a subsurface cone of water to each plant roots. As with using furrows/trenches, allows the water to penetrate deeply, and water only when the soil surface is completely dry. The disadvantage of flood irrigation is that it required excess irrigation water and also leads to increase in insect-pest and diseases.

Fig. 3.3: Flood Irrigation Method for Home Garden

v) Drip irrigation: It is also known as trickle irrigation. Application of water to the soil in this is through small orifices or emitters which are designed to discharge water at rates of 1-8 liters/hour. The emitters are installed close to plant, wetting only specific areas, leaving the rest of the field dry – unlike sprinkler irrigation and flooding which wet entire field. Water is therefore, used efficiency.

Advantages: Following are its advantages.

- Sanitation – The foliage is kept dry and spread of soil diseases and weed seeds through surface flow of water is barred. With the controlled release of water, the soil is not waterlogged, a condition which may favour development of some diseases.

- Flexibility in farm operations – Cultivation, spraying, and harvesting can be done even while irrigating because field is kept dry, except area close to plant.

- Uniform water distribution

- Ease in combining irrigation with fertilizer and pesticide application.

- Savings in terms of labour (for irrigation, weeding, fertilizer and pesticide application), water, fertilizer, and pesticides.

- Drip irrigation system: From the pump, water flows to filtration system, which removes sand, slit, weed seeds, and other foreign matter. The water is then evenly distributed to laterals and flows through emitters. The pressure of the sub-main line is controlled by pressure regulator at preset pressure appropriate to the system. Fertilizer and pesticide injectors are installed optionally before the filter.

vi) Sub-irrigation: It is the least common method of irrigation because of its high initial cost and limited land suitability (usually peat). Water is supplied by an underground system and reaches the plant by capillary movement. It allows precise application of water, nutrients and other agro- chemicals directly to root zone of plants. Drip irrigation system can be installed underground to serve as sub- irrigation. Conditions favouring use of sub-irrigation include:(i) Soil must permit rapid lateral and downward movement of water. ;(ii)The topography of the land should be smooth, uniform, and approximately parallel to the water table.

2.25.9 Water Quality

Properties of water making it suitability for irrigation are as follows.

- salinity which can be measured by electrical conductivity (EC) or salt content,
- alkalinity which is expressed in terms of sodium concentration, and
- toxic components, specially boron standards

Table 2.4: Three Classes of Water Based on These Parameters are as Follows.

Water Class	Electrical Conductivity (micromhos/cm)	Salt Content (ppm)	Sodium (%)	Boron
1	0-1000	0-700	60	0.0 – 0.5
2	1000-3000	700-2000	60-75	0.5 -2.0
3	0ver 3000	Over 2000	75	Over 2.0

Class 1 water is considered excellent to good, suitable for most plants under most conditions.

Class 2 water is considered good to injurious, harmful to more sensitive crops.

Class 3 water is considered unsuitable under most conditions and harmful to most crops.

2.25.10 Mulching

It is a cultural practice that can significantly decrease the amount of water need to add to the soil. A 5-7.5 cm (15-20 cm of loose and transpire moisture straw or leaves will compact to 5 to 7.5 cm of mulch) organic mulch can reduce water needs by as much as half. Mulches smother weeds, which take up and transpire moisture, and reduce the evaporation of moisture directly from the soil. Organic mulches themselves hold some water and increase the humidity level around the plant. If the mulch becomes dry, it may be necessary to add an extra 2.5-5 cm of water to soak through the mulch when doing overhead watering.Black plastic mulch also conserve moisture, but may increase soil temperatures dramatically in summer (to the detriment of some plants and the benefit of others) if not covered by other much materials of foliage.

2.25.11 Water Reuse for Home Gardens

Water reuse can be defined as the use of reclaimed water for a direct beneficial purpose. It has been employed as a water conservation practice in many countries. Reclaimed water, also known as recycled water, is water recovered from domestic, municipal, and industrial wastewater treatment plants that has been treated to standards that allow safe reuse.

Gray water, or untreated waste water from bathing or washing, is one form of wastewater. Places where the use of gray water is allowed, the following rules are recommended:

-Apply gray water to the soil, not to plant leaves.

-Do not use "black water" (any water run through the toilet) because of the possibility of contamination from fecal organisms.

-It is best not to use kitchen wastewater that contains grease, harsh cleaners, ammonia, bleach, softeners, or non-biodegradable detergents.

-If using water from the bathtub or washing machine, use only mild, biodegradable soaps. Omit softener sand beaches. Allow wash and rinse water to mix, if possible, to dilute the soap content. Never use a borax-containing product (such as washing soda) in water to be used on a garden because of the danger of applying toxic levels of boron.

Excess water, whether it comes from irrigation or from natural source, must be removed from the field to ensure normal crop growth. Poorly drained soil does not only harm the crop directly, but also causes problems with scheduling of farm operations and accelerates disease development.

Drainage Methods and Applications

Surface drainage: This method eliminates ponding, prevents prolonged saturation, and accelerates flow to an outlet without siltation or erosion of soil. In some cases, orientation of two crops with the land slope may accomplish this purpose. In other cases, the use of diversion is necessary. Surface drainage systems include both collection and disposal ditches. The cross section, slope, pattern, and small spacing of ditches are essential factors of design where system, or parts of the system, primarily collects and removes surface water from a field or small land area. Ditches for surface drainage are usually designed to remove runoff, produced by an ordinary rain to prevent damage to crops grown in the drainage area.

Subsurface drainage: This is defined as the removal of excess groundwater. Surface ditches are necessary to remove excess runoff from precipitation and dispose surface flow from irrigation; but these should be designed to complement subsurface drainage system.

2.26 Water Conserving Techniques

2.26.1 Amendment of Garden Soil

In the vegetable garden, the routine addition of organic soil amendments such as compost would optimize potential yields and produce quality. On sandy soils, organic matter holds over ten times more water and nutrients than the sand. On clayey soil, organic matter glues tiny soil particles together into larger aggregates, increasing pore space. This increases soil oxygen levels and improves soil drainage, which in turn increases rooting depth allowing roots to readily reach a larger supply of water and nutrients. Organic matter also encourages beneficial activity of soil organisms and helps remediate soil compaction. To reduce soil compaction and help conserve moisture in the soil during the winter, cover newly cultivated garden with mulch.

2.26.2 Reducing Water Need

Choose the irrigation system that is most efficient for the needs; Use of a drip system on a mulched garden reduces water needs by around 50%. With a drip system, water timers added to the hose at the faucet reduce overwatering. Drip or soaker hoses and micro-spray systems are good for areas which dry -out quickly. As part of an efficient irrigation system, check soil moisture regularly. Squeeze soil in your hand; if it sticks together, it is moist and should not be irrigated. If it does not stick together, then it is time to irrigate. Irrigate in the morning when temperature is cool but rising.

2.26.3 Other Water Saving Techniques

This creates shade for roots and reduces evaporation; Use raised bed techniques that do not allow water to flood and reduce water stagnation and weed germination as well; Control weeds that compete with vegetables for water; Group with similar water needs in the same section of the garden for easy irrigation. Cucumber, zucchinis, and squash, for example, require similar water applications; **Protect plants and soil from wind** with windbreaks to reduce evaporation;Position sprinklers so you're not watering the side of the house, sidewalk or street;Apply water slow enough so that run-off doesn't occur; One deep watering to fill root zone with water is much better than watering several times lightly; Watch your plants: They'll let you know when they need water: They wilt. Colours become dull. Footprints in the lawn stay compressed for more than a few seconds;Buy and install rain barrels. These help reduce runoff and collect rain water for plants and outdoor uses.

2.27 Critical Water Periods for Vegetables

As a rule of thumb, water is most critical during seed germination, the first few weeks of development, immediately after transplanting, and during flowering and fruit production. Plant age also matters; young plants should be watered immediately after transplanting and mature ones require less water than plants in the middle of their growth cycle. The critical watering periods for selected vegetables are as follows.

i)   Beans have the highest water use than any common garden vegetable. They use 0.6 to over 1.25 cm inch of water per day (depending on the temperature and wind) during blossoming and fruit development. Blossoms drop with inadequate moisture levels and pods fail to fill. On hot, windy days, blossom drop is common. When moisture levels are adequate bean plant is a bright, dark, grass-green. As plants experience water stress, leaf colour takes on to a slight greyish cast. Water is needed at this point to save blossom from drop.

ii)   Carrot and other root crops require consistent moisture. Cracking, knobby and hot flavoured root crops are symptoms of water stress.

Cole crops (broccoli, cabbage, cauliflower, collards, Brussels sprouts, kale, and kohlrabi) need consistent moisture during their entire life -span. The quality of cole crops is significantly reduced if the plants get dry anytime during growing season. Water use is highest and most critical during head development.

iii)   Corn water demand peaks during tasseling, silking, and ear development. Water stress delays silking period, but not tasseling. Under mild water stress, crop may tassel and shed pollen before silks on ears are ready for pollination. The lack of pollination may result in missing rows of kernels, reduced yield, or even eliminate ear production. Yield is directly related to quantity of water, nitrogen and spacing.

iv)   Lettuce and other leafy vegetables need water most critically during head (leaf) development. For quality produce, these crops require a constant supply of moisture.

v)   Onion family crops require consistent moisture and frequent irrigation due to their small, inefficient root system.

vi)   Peas need water most critically during pod filling.

vii)   Potato tubers will be knobby if they become overly dry during tuber development.

viii) Tomato family (tomatoes, peppers and eggplant) needs water most critically during flowering and fruiting. Blossom end- rot (a black sunken area on the bottom of the fruit) is often a symptom of too much or too little water. The tomato family has a lower water requirement than many vegetables; and plants are often over-watered in the typical home garden.

ix)   Vine crops: cucumbers, summer and winter squash, and assorted melons need water most critically during flowering and fruiting. Vine crops use less water than many vegetables and are often over-watered in the typical home garden.

2.28 Watering Timings and Techniques

Timing of watering is important, evening is the best time to irrigate or drip irrigate –this gives soil whole night to absorb water. Early morning is the best time for sprinklers – leaves can absorb water and not to be wet and cold all night. Loosen soil around vegetables so that the soil can quickly absorb water. Apply water in furrows or basins around the crops to reduce evaporation losses – dig furrows between crops rows about 10-15- cm deep. Place mulch between crop rows to reduce evaporation. Use small amounts at a time to avoid mould or root problems. Soak vegetable garden once a week to a depth of 15-30 cm and don't water again until the top few centimeters begin to dry out. Eliminate weeds. Plastic mulch around vegetables not only saves water, but it promotes early plant growth and cuts down on weed establishment. Plant tomatoes, cucumbers, and squash in hills and group them up whenever possible. Raised beds are good to conserve water and space.

There are summers when the rainfall is less and the heat abundant, making it difficult to have a successful garden. Some tips to help with that situation include the following.

- Use mulches to help ground hold water better; besides helping to save moisture, mulches also prevent weeds from growing.

- Harvest young vegetables: When picked young, vegetables don't use up as much water in the long run. They usually taste better too. Some crops, such as leaf lettuce even produce more if harvested often.

- Pull weeds: Vegetable plants would have to compete with weeds for every drop of water so weeds should be out as soon as seen. If using a hoe, do not dig too deeply.

2.29 Direction

Proper sunlight on vegetable garden ensures healthiest and highest crop yields. Vegetable gardens should face southward. A south-facing garden receives the largest amount of sunlight as the sun passes overhead throughout the day. If possible, plant garden on the south side of the house. If that's not possible, the second best option is to situate the garden on the east or west side of house, as this direction also receives a large amount of light. If possible plant vegetables in rows from north to south directions. As by this method the whole plant will receive sunlight once on the whole plant throughout the year. A garden on the north side of the house is least desirable. Surrounding homes, buildings, trees and shrubs cast shadows over the garden for the majority of the day in a north-facing garden.

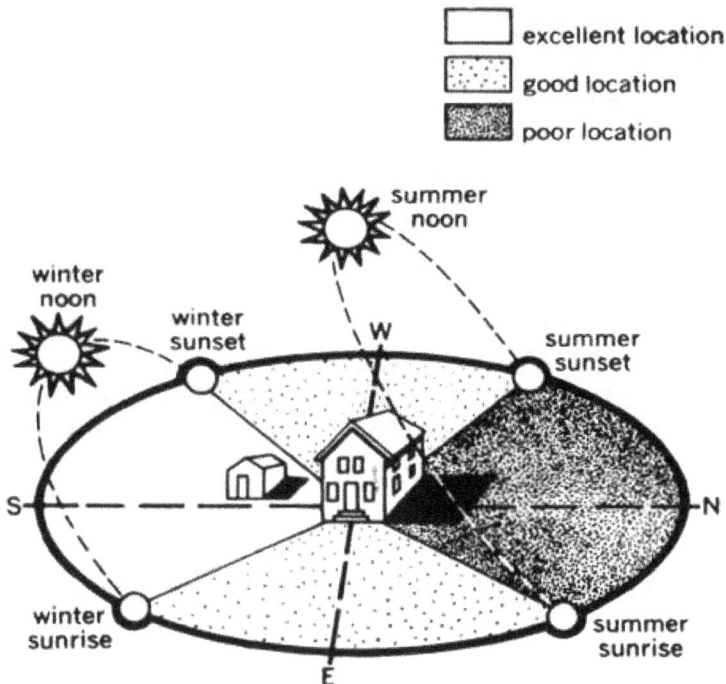

Vegetable gardening is a rewarding, relaxing way to enjoy the outdoors and get some light exercise. Also reap a harvest of fresh, safe, healthy vegetables, at a fraction of the cost. To get the most out of the garden, however, supply your plants with optimal growing conditions, including full sun. The direction of home garden plot faces would play a role in the amount of sunlight your plants receive throughout the day.

Plants require sunlight for photosynthesis, to convert the sun's light into food and energy. Most vegetable plants are full-sun plants, requiring between six and eight hours of sunlight per day. Choose a plot that receives good morning sunlight and receives dappled or partial shade in the afternoons.

2.30 Layout

Planning a Vegetable Garden Layout: Step-by-Step Success

- Planting row should be from north to south direction.
- Plant tall crops that cast shade, such as beans and corn, in the northern half of the vegetable garden.
- Group perennial vegetables together, preferably along the edges.
- A small hedge placed around the site can provide protection from the wind.
- Consider sunlight, soil, and climate when locating the ideal garden spot.

- Cold air drains downhill and tends to collect in low-lying areas.

- The coldest and most frost prone location is at the bottom of the hill.

- The warmest site is always at the top of the slope.

An attractive and productive vegetable garden is a source of pride that also provides a bounty of fresh wholesome produce. A wealth of pleasure awaits you while implementing a well-designed garden plan.

When you are ready to add plants to the vegetable garden, make sure your rows run north-south. This optimizes the amount of light they will receive. It is also important that you allow the proper amount of space between plants in your garden. Plant spacing is dependent on the type of vegetables you are growing. Take note of the taller plants that you plan to grow and ensure that they do not cast shadows over smaller plants in your garden

When planting a garden, it is a must to consider structures and elements in the vicinity that would cast shadows over the vegetable plot. Tall buildings, other homes, trees and shrubs play a part in creating shade over your garden at various times of the day. This is helpful during the afternoon, when partial shade is desired, but a hindrance if it blocks morning sun. As a rule, never plant your garden next to a tall building or under a shade -tree, since these locations interfere with full sunlight.

2.31 Implementing Kitchen/Home Garden Design

The designed kitchen/home garden model on 6m x 6 m area each incorporates easy to grow, nutritious and indigenous vegetables. This area could be easily handled by household women and the produce could be consumed at least by a 4-member household. This area can be divided into five longitudinal blocks which are further sub-divided into 2-3 smaller sub-plots measuring 2mx1m, 3mx1m and 1mx1m, respectively, depending upon the crop. Farmyard manure should be mixed in the plot area, sowing and transplanting of vegetable crops should be done based on the crop season and duration. Fertilizers and management practices should be applied as recommended. The crop selection should be based on the location specificity, cropping seasons, nutritional availability, performance and family preference. In this model, 27 crops can be fitted into 13 cropping sequences in an example for North India. (W). The model is estimated to produce 250-280 kg of fresh vegetables in a year.

i) **Total edible yield:** Vegetables can be harvested from kitchen gardens twice or thrice every week, depending upon their stage of maturity providing average daily yields of about 200 g per person in a four-member household under North Indian conditions. The home -garden design in 6m x 6m for north India could significantly improve household access to vegetables and is sufficient to meet recommended levels in most times of the year.

ii) Nutritional yield: The nutrient yield data suggested that while the 6m x 6 m vegetable gardens achieved supplies at recommended vegetable consumption level, vitamin A and vitamin C supplies were sufficient and continue to provide 100% - 500% of RDA, but protein and iron supplies were difficult to achieve (30% of RDA). High protein legumes and very high iron vegetables should be included in the garden designs to improve protein and iron supplies.

Table 2.4: Cropping Sequences for Indian Home Gardens Under Plain Conditions

Plot Number	First Crop	Follow-up Crop1	Follow-up Crop 2
1	Bottle gourd(Jun.- Dec.)	Onion(Jan- May)	-
2	Radish (Jul.- Sept.)	Garlic (Oct-Apr.)	Lettuce (May-Jun.)
3	Coriander (Jul.- Jun.)	-	-
4	Eggplant(Jul.-Dec.)	Lablab (Sept.-Feb.)	
5	Chilli(Mar.-Oct.)	Kasurimethi(Nov.-Feb.)	-
6	Amaranthus (Jul.-Sep.)	Spinach (Oct.-Feb.)	Cowpea (Mar.-Jun.)
7	Sponge gourd (Jul.-Nov.)	Tomato (Dec.-May)	-
8	Mint (Jul.-Oct.)	Chinese cabbage (Nov.-Feb.)	Long melon (Mar.-Jun.)
9	Kangkong(Mar.-Nov.)	Veg. mustard (Nov. Feb)	-
10	Cowpea (Jul.-Oct.)	Garden pea(Nov.-Feb.)	Okra (Mar.-Jun.)
11	Basella (Mar.-Oct.)	Broccoli (Nov.-Feb.)	-
12	Okra (Jul.-Oct.)	Carrot (Oct.-Feb.)	Cucumber (Mar.-Jun.)
13	Spinach (Jul.-Jan.)	Capsicum (Feb.-Jun.)	-

Home garden model: Andhra Pradesh

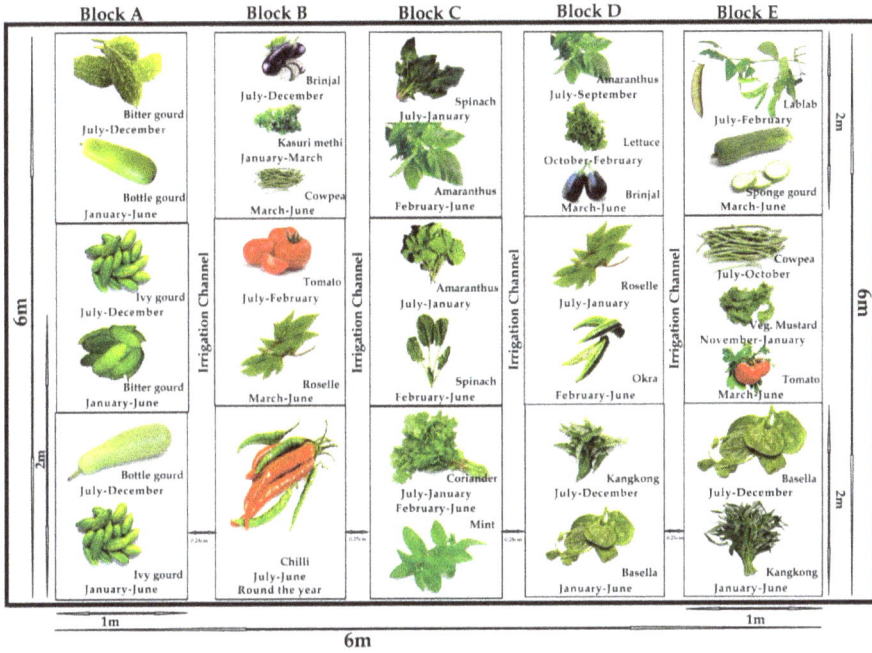

Fig. 3.4: Home Garden Design for Andhra Pradesh

Home garden model: Jharkhand

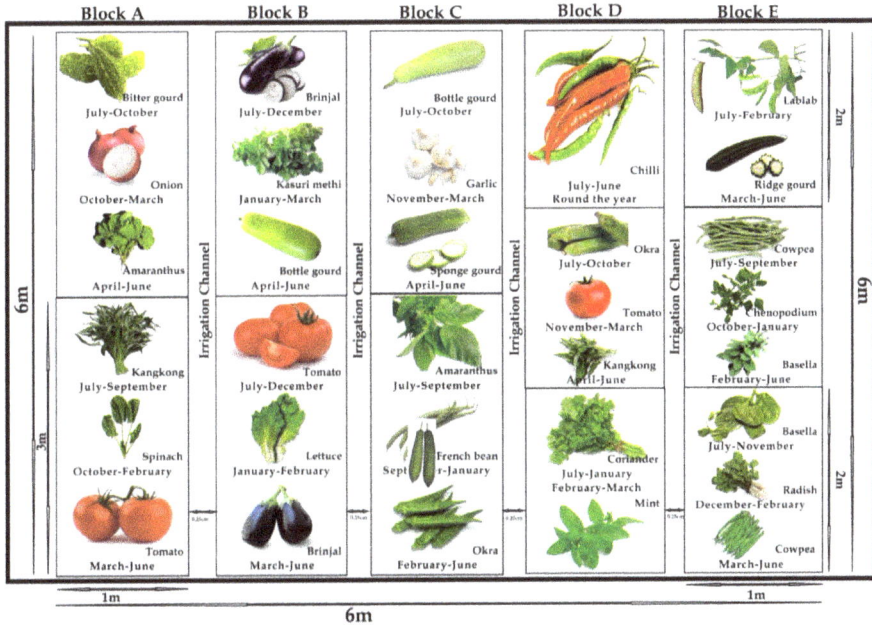

Fig. 3.5: Home Garden Design for Jharkhand

Home garden model: Punjab

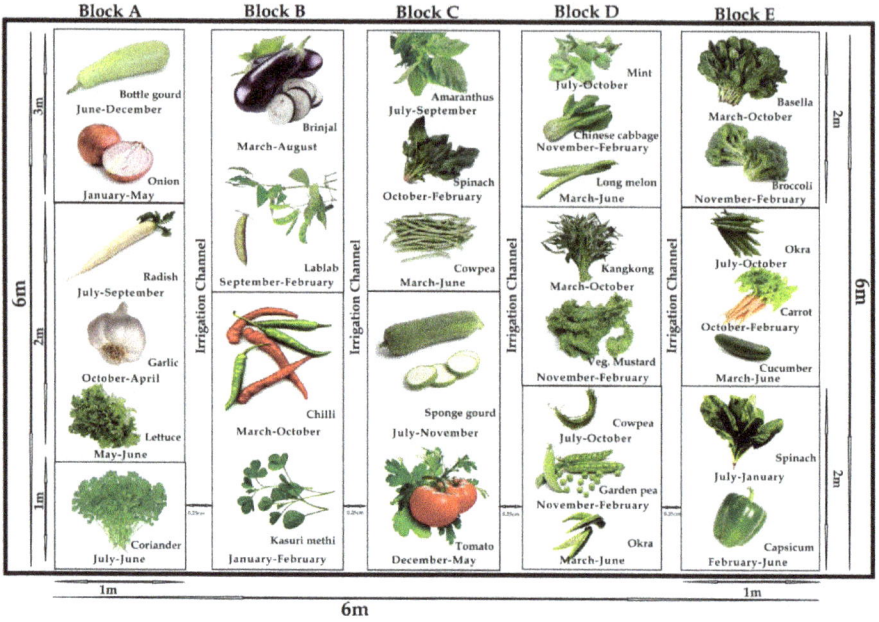

Fig. 3.6: Home Garden Design for Punjab

3

Types of Gardens

The domestic garden can assume almost any identity within the limits of climate, materials and means. The size of the plot is one of the main factors, deciding not only the scope but also the kind of display and usage. Limits on space near the urban centres as well as the desire to spend less time on upkeep have tended to make modern gardens smaller. No matter if they are urban or rural, all gardens are benefitted from pre-planning and design.

Gardens can be as follows: **Home/nutritive/kitchen gardens; School gardens; Hanging/vertical gardens/ terrace gardens; Community gardens; Hydroponics; Urban gardens**

3.1 Home Gardens

They are also known as nutritive, kitchen or even vegetable gardens. There are two types of home gardens:

i) Fixed and ii) Scattered.

In a fixed home garden, land is compact. It can be divided into different sections and beds as desired.

3.1.1 Types

i) In-ground Garden Beds

An in-ground garden of 500 square feet is enough to feed a family of four for eight months in a year with plenty left to share with family, friends, and neighbours. Inputs needed are hand tiller, seeds, compost, trellis and stakes to properly grow tomatoes and squash and small fence. The most common vegetables and fruits grown successfully in these gardens are: Cabbage, corn, beets, pumpkins, watermelon, squash, tomatoes, peppers, potatoes, eggplant. In these beds, they are easiest to be established; nothing is to be build; ; pathway weeds can creep onto the bed ; more likely pets and children would walk on the beds; and you have to reach further down to tend plants

Fig. 3.1

ii) Sunken Raised Garden Beds

These are prepared by digging down the pathway topsoil and adding it to beds; this is a way to fill raised beds without importing soil. These have same benefits as raised beds, but less drainage in wet months. It has less visual impact in the yard. While the beds are 12" deep, surface of the beds is only 4" - 6" above the ground level (Fig. 3.1).

iii) Raised-bed Garden

A raised-bed garden (Fig. 3.2) gives you a little more space to experiment and will offer higher yields. This is also ideal for a slightly larger area, and can be a great alternative to direct-to-ground planting, especially if the soil conditions are not ideal. For this horticultural adventure, all needed is cut cedar boards (wood is rot-resistant), a couple of layers of newspapers, top soil, peat moss, compost, grass clippings, seeds, water and a hand tiller.

Fig. 3.2

Plants that grow well in the raised-bed gardens are: Rhubarb, asparagus, beans, broccoli, brussels, pumpkins, spinach. These garden beds provide best drainage and doesn't let soil compaction; soil warms up more quickly in spring; bed sides prevent weeds from creeping into the bed; in them it is easier to tend plants because soil level is raised; and not the least serve as a barrier to pests such as slugs and snails.

iv) Window-box Garden

Even if someone lives in an apartment, still horticulture can be tried with a window- box garden. Inputs required are window box or hanging box (to hang off your deck), garden soil, seeds and water. The best type of vegetables to grow in a window- box garden are lettuce, greens and spinach, as their seeds are sown on the top of the soil, and they don't have deep roots. To grow tomatoes, carrots and tubers, either deeper boxes are built or bought. Spices such as oregano, basil and chives also do quite well in this type of garden.

ii. A scattered garden consists of plots in different areas. Many families do not have a fixed piece of land. Instead, they have various small areas around the home. Use of these small areas to grow different vegetables constitutes a scattered garden. Scattered gardens are more familiar than fixed gardens to landless and marginal population.

3.1.2 Characteristics of a Home Garden

wet near ring wells and tube- wells, less moist or dry by roadside, shady under trees, lowland and highland with good sunlight. These areas can be used to grow different vegetables, and would ensure a varied supply of vegetables throughout the year.Fig. 3.3)

- Vegetables like mint can be grown near ring- wells and tube- wells

- Fruit vegetables like tomato, brinjal and okra, requiring full sunlight, can be grown in exposed areas

- Climbing vegetables including gourds, drought- resistant vegetables, including kangkong and legumes and shade- loving vegetables like amaranth and other green leafy veggies can be grown close to house along the roadsides and under the shade of other trees.

Fig. 3.3

3.1.3 Advantages of Home Garden

Following are the main advantages.

- Produces a steady supply of vegetables year- round.

- Vegetables can be harvested at optimum maturity and eaten or preserved while fresh.

- Fresh vegetables are safe, higher in flavour, nutritive value and lower in cost than purchased vegetables, which may also be full of pesticides.

- Home gardening is a healthy exercise and an interesting outdoor activity for the family.

- Give a feeling of accomplishment, self-sufficiency and security.

- Saves on daily money spend on vegetable purchase and also adds on income to households from excess produce.

3.2 School Gardens

Fig. 3.4

A school garden (Fig. 3.4) can be an area of land within the school ground or nearby to it for growing vegetables, fruits, flowers, medicinal plants, trees, bushes and many other plants. In cities where schools have limited space or lack open spaces, school garden can be the plants grown in containers. Schools with vegetable gardens produce healthier young people with healthier attitudes to life. Garden-based learning is hands-on, minds-on experience where both students and teachers can learn together.

Adding gardens in all shapes and sizes to the school yards serve as tools for outdoor science and environmental education. These gardens have been observed to increase children's respect for nature and to aid in physical, mental and moral development.

School gardens usually have the following two things in common : (i) The schoolchildren actively help parents and other interested community members in creating and maintaining the garden ; (ii) The schoolchildren use the garden for learning, for recreation and also eat fresh what is harvested.

3.2.1 School Gardens and Nutrition Education

School gardens provide a site for hands-on learning and to practice using scientific method. Besides, children are excited about eating fruits and vegetables which they grew themselves. Students would have greater appreciation for how the food is grown. Gardens can be used to teach proper post-harvest management and food safety through proper harvest, processing and storage. Children may have the opportunity to practice preparing nutritious foods and to try new foods in their diets. Gardening is a skill which children can use for the rest of their lives to promote health and wellness.

Nutrition education through school gardens increases children's knowledge about fruits and vegetables, which may improve their attitude towards the food and may lead to have adoption of better eating habits.

3.2.2 School Gardens and School-Feeding Programmes

School feeding is a powerful tool to alleviate short-term hunger and enhance children's learning capacities. It also provides an incentive for parents to send children to school, particularly girls. School gardens, if planned and implemented with the support of parents and the community, can complement school feeding programmes and enhance their long-term impact in terms of children's health/nutritional status and learning abilities. The promotion of micronutrient-rich vegetables, including indigenous vegetables and fruits in school, home and community gardens would diversify local food base, generate income and add nutritional value to children's school meals, thus contributing to their nutrition.

3.2.3 Pointers for School Gardens

- Class plots/ experiments should be large, not less than 400 square feet (20ft × 22 ft)
- The simple rectangular form of plot is most desirable; intricate geometrical forms and designs should be avoided.
- Individual plots for third and fourth class students should not be less than 5 ft × 8 ft, nor more than 6ft × 10 ft, and for first and second classes, about 4 ft ×6 ft.
- All growing plants and bulky crops, corn, potatoes, tomatoes, and vine crops (cucumbers etc.), should not be placed in individual plots, but in large class plots.
- Care must be exercised by teachers in choosing varieties of flowers and vegetables.
- A student should not be allowed to grow more than 2 or 3 varieties of flowers and vegetables at once.
- Early crops should be followed by late maturing crops, such as tomatoes, cabbage, beetroot, turnip etc.

3.2.4 Advantages of School Garden

Following are the advantages of school gardens.

- Promoting good nutrition and addressing multiple learning styles.
- Improving environment and increasing physical activity.
- Teaching patience and responsibility and teaching students to work with cooperation.

- Improving social skills and building classroom relationships and school spirit.
- Supporting interdisciplinary education and creating awareness for team-work.
- Increasing self-esteem and self-worth and beautifying school environment.

3.3 Community Gardens

A community garden is a single piece of land gardened collectively by a group of people.These are managed and maintained with the active participation of the gardeners themselves. Such gardens encourage food production by providing gardeners a place to grow vegetables and other crops. To facilitate this, a community garden may be divided into individual plots, depending on the size and quality of a garden and the members involved. Some gardens are grown collectively, with everyone working together; others are split into clearly divided plots, each managed by a different gardener (or group or family). These gardens provide lot of benefits.

Fig. 3.5: A Community garden in Kadakampally, Kerala

3.3.1 Individual Benefits

- Community gardening is an active pursuit yielding fresh foods to individuals; and families have access to fresh, nutritious food and mixed meals supporting nutritional health and promote physical fitness.
- Health benefits to be associated with involvement in community gardening include decreased stress, increased overall sense of well-being and reduced risk of childhood lead poisoning.

- Eating locally produced food reduces asthma rates, because children are able to consume manageable amounts of local pollen and thus develop immunity.

Learning to grow plants is mentally stimulating and adds to individual's knowledge and expertise. Gardens are used for community education such as waste minimization and recycling of wastes through composting and mulching.

Social Benefits: Community gardening (Fig.6) is a social activity involving shared decision- making, problem -solving and negotiation; increase these skills among gardeners. People come together with a common purpose as community gardens are places where people meet others from a wide variety of backgrounds (age, race, culture, social class). These also add beauty to community and heighten people's awareness and appreciation for living things. These can be used to build a sense of community and belonging.

Community gardens offer unique opportunities to produce traditional crops otherwise not available locally and provide inter-generational exposure to cultural traditions and offer cultural exchange with other gardeners. Community gardens allow people from diverse backgrounds to work side-by-side on common goals even if not speaking same language.

3.3.2 Improving Urban Environment

Community gardens re-green vacant plots and bring vegetational diversity to public open space and other areas, making them a useful tool for urban improvement. By diversifying use of open space and creating opportunity for passive and active recreation, community gardens improve the urban environment. The diversity of plant types found in community gardens provides habitat for urban wildlife, increasing their value for improving the natural environment. They provide a place to retreat from noise and commotion of urban environments as well as much needed green space in lower-income neighborhoods.

3.3.3 Economic Benefits

Community gardens can be a significant source of food and/or income for community members. These are especially helpful for families and individuals without much land who would not otherwise be able to produce their own food. Urban agriculture is 3 to 5 times more productive per acre than traditional large-scale farming. Studies show that community gardens can increase neighborhood property values.

3.3.4 Youth Engagement

Community gardening is a healthy, inexpensive activity for youth which can teach them about appreciation for the nature and how to interact with others in a socially meaningful and physically productive way. Not only can youth gain practical job and life skills through gardening, they can also learn about the work that goes into getting the food they eat and about the importance of community stewardship and environmental sustainability.

3.3.5 Establishing a Community Garden: Some Pre-requisites and Steps

Is there sufficient interest? The initiative must come from individuals who actually will be using the garden.

Where will the garden be located? One of the first practical tasks will be to secure a usable plot of land. The area to be used as a garden should have a reliable source of water.

3.3.6 Advantages of Community Gardens

They provide access to fresh produce and plants as well as access to satisfying labour, neighborhood improvement, sense of community and connection to the environment. They are publicly functioning in terms of ownership, access, and management, as well as typically owned in trust by local governments on nonprofits. A community garden brings community closer. These gardens encourage organization for an urban community's food security, allowing citizens to grow their own food or for others to donate what they have grown.

3.4 Hanging/Vertical Gardens

Vertical gardening in its simplest form is the idea of vegetables that in their natural state may tend to sprawl over ground, and provided with a support structure for allowing them to grow vertically. A vertical garden is very practical for anyone with a small space to work. Growing vegetables upright also makes easy harvesting.

Fig. 3.6

Vegetables that can do well in a vertical setting (Fig.7) include climbing plants like beans, peas. Tomatoes are often staked; indeterminate varieties like cherry tomatoes often a favourite. Nontraditional choice for vertical gardening is vining plants such as melons, squash, cucumber and pumpkin. These are a lot like grapes, their vines produce tendrils that wrap around structures and climb as they grow.

3.4.1 Support Structure for Vertical Gardens

- Simple stakes or poles: Often used for peas and beans. Simple trellis: A common choice for cucumber. When placing a trellis or stake, be aware of your seasonal wind patterns, and be sure that mature plant is adequately supported.

- Fence: If there is enough sunlight (perhaps a southern exposure) one can use nails to string fishing- line or weather-proof twine or other support netting on a wooden fence, or a fence built of chicken -wire or some similar structure to provide support.

- Support cage from wood or PVC pipe: ombined with netting to support larger fruits. This is a good choice for some of the cucurbits like melons or pumpkins

3.4.2 Setting-up a Vertical Vegetable Garden

First is to determine what are the conditions, like the area one wishes to set-up vegetable garden, such as in the balcony. The amount of sunlight would be a greatest factor in determining which plants would thrive in the urban environment. If the area is surrounded by other buildings, balcony or patio may be shaded most of the time; therefore, one should choose the plants accordingly. Leafy vegetables like lettuce, cabbage, and greens do well with limited sunlight, making a good choice for shady areas. If there is abundance of sunshine, the selection of plants would be more, as vegetables thrive best in full sun. Choices here can include tomatoes, peppers, potato, beans, carrots and radish. Even vine crops, such as squash, pumpkins, and cucumbers, can be grown as long as the container is deep enough to accommodate them and proper staking is available. Fill containers with peat moss and a suitable potting mix amended with compost or manure. Steps are as follows.

- Pick the vegetables that you want to grow. There are many that will work with vertical gardens. Beans are great climbers to use. Corn stalks can be positioned against gardening poles.

- Pumpkins can be used at the base for good mulching ground coverage. The pole beans can even climb up the poles that the corn stalks are positioned against or simply up corn stalks. Winter squash can be grown indoors in little peat pots. When these grow, use garden string to tie vines to the trellis.

> ➤ Soft cloth or gardening ropes can be used to prevent any damage to vines.

> ➤ Vining cucumbers can be grown up the trellis using thick garden string to hold vines to the boards. Garden cages are also good for this. Snow peas can be grown up the trellis, garden cage or netting.

- Use sturdy wire, metal or wood supports to grow any type of tomato vertically. Use stretchable green tape for securing tomato- plants to the structure they are growing up against.

- Water the vertical garden frequently as the plants will be up in the air, they will dry out quickly. On the other hand, they are less likely to have fungal issues.

> ➤ For any seeds that are sown, make sure ground never becomes completely dry.

> ➤ Old plastic grocery bags can also be used.

- -Punch tiny holes in the bottom; these retain moisture well and good results have been found with many different types of vegetables.

- -Hanging baskets can be placed on the balcony or on suitable hangers. Numerous types of vegetables can be grown in hanging baskets, especially those with trailing characteristics. Water them daily, since hanging baskets are more prone to drying out, especially during hot spells. Trellises can be used for support of trailing or vine crops. A stepladder can be used as a makeshift trellis to support vine-growing plants like pumpkins, squash and cucumbers. The rungs of the ladder can be used to train vines while placing vegetables on its steps for further support;-this also works well with tomato- plants. Growing a vertical vegetable garden is the perfect way for urban gardeners and others to still enjoy a bountiful harvest of freshly grown vegetables.

3.4.3 Low-tech Version Vertical Gardens

Construction debris bags with poked holes for lettuce and other garden veggies to peek out of, form a great way to build a garden. This requires a lot less weeding.

3.4.4 Advantages of Vertical Gardens

Can make use of some otherwise unusable space for the garden. Areas like rooftops or patios can add to the garden with this approach. A vertically trained plant can be used to improve aesthetics of the yard; hiding things like utility transformers or boxes with a trellis or small fence, or simply dress up a boring wall.

3.5 Container/Terrace Gardens

Container gardens are very easy to set up and get started. The only supplies needed are Containers; Gardening soil;Hand rake or tiller;Seeds; Water;Compost (for heavy-feeding plants like squash); and Shade -nets in hot and arid areas during summer

3.5.1 What to Grow ?

The most important thing to remember when planting in containers is that the roots of the plants can only go down so far. Make sure that the containers are deep and wide enough to accommodate vegetables. For example, most of vegetable containers are approximately 12 to 14 inches wide and 10 to 12 inches deep. Here is a list of 10 vegetables that grow really well in containers: Tomatoes; potatoes; cucumbers; carrots; peppers; green onions; turnips; green beans; lettuce; and squash.

While carrots and tomatoes grow well together in the same container, squash needs to be grown in separate pot; its a heavy-feeder that needs lots of compost, which other plants do not. As its vines grow, it can choke out other plants and keep them from moving past seedling point.

Main considerations for a container garden are as follows: Choosing a proper container; Using a good soil mix; Variety selection; Planting and spacing requirements; Fertilizing; Watering; Providing 5 hours or more of full sun

3.5.2 Choosing a Proper Container

Type of the container depends on the vegetables grown. Containers can be made of clay, wood, plastic or ceramic. Wooden barrels, decorative boxes, plastic garbage cans, tin cans, plastic laundry baskets etc can be used. Small containers dry out quickly and may blow over in the wind. Square, rectangular or circular containers work equally well.

 Dark coloured containers should be avoided because they absorb heat which could possibly damage plant roots. If dark coloured pots are to be used, try painting them in a lighter colour or just place the container in shade.

Container should have adequate number of holes in the bottom for proper drainage. Additional holes are drilled or punched in containers (Fig.8) that do not drain quickly after each watering. If the container does not have drainage holes, bottom 1/4 of the container can be filled with rocks or pebbles to hold excess of water until it evaporates or is used. Container should be raised 2.5-5 inches off the floor by setting on wooden blocks for proper drainage. Plastic materials are non-porous and retain more water; plants grown in these containers dry out less rapidly and can be watered less frequently.

3.5.3 Soil Mix/Growing Medium

Growing medium has three main functions— supply roots with nutrients, air, and water; allow for maximum root growth; and physically support plant.

Best to use are commercial potting mixes. These are light weight, fast draining and free of insects, diseases and weeds. Peat-based mixes, containing peat and vermiculite, are excellent. Light weight mixes though good for hanging baskets, window boxes and containers that are moved around but are not suitable for growing large plants such as

Fig. 3.7

sweet- corn, staked tomatoes and eggplant. The home made mixes are made from equal amounts of good garden soil, washed coarse sand and organic material such as peat moss, leaf mold or sawdust. The mixes need to be free of various pests, and this can be achieved by heating them at a low temperature in the oven. Compost is dark, crumbly, earthy-smelling product of organic matter decomposition. Leaves, grass clippings, wood waste, and farm animal manures are some of the common ingredients that are combined with water in piles or windrows and digested by huge populations of oxygen- loving microorganisms.

- **Some good media mixtures for container vegetables are:** 100% compost;100% soil-less mix; 25% garden soil + 75% compost; 25% soil-less mix + 25% garden soil + 50% compost; 25% garden soil + 75% soil-less mix; 50% soil-less mix + 50% compost

- **Recommended media depth are:** *10-15 cm:* salad greens, Asian greens, mustards, garlic, radish, basil, cilantro, thyme, mint, and marjoram. (Salad greens and some herbs have shallow, fibrous root systems, and are well suited to shallow containers with a large surface area); *20-30 cm:* beans, beets, chard, carrots, cabbage, pepper, eggplant, tomato, squash, rosemary, parsley, lavender, and fennel.

- **Vegetables for containers:** A number of varieties have been designed for small gardens and container gardens: These varieties do not grow large but still produce good yields: Most herbs, parsley require a standard 15- cm pot; beetroot, lettuce, onions, carrot and radish require a container holding about one gallon of soil mix; pepper and small tomato varieties grow best with two gallons of soil; Large vegetables like cucumbers, eggplants, sweet -corn and tomatoes grow best in four gallon container or even larger.

- **Required pot volume:** *1-3 gallons:* herbs, green onions, radishes, onion, chard, pepper, dwarf tomato or cucumber, basil; *4-5 gallons:* full-size tomatoes, cucumber, eggplant, beans, peas, cabbage and broccoli.

- **General care:** Plants in containers should not be crowded. Thinning should be practiced, to allow ample growing space for plants. Root crops and greens should be planted on the basis of the space needed when mature, not as seedlings. Don't cram medium into container. Fill to within 2.5 cm or so of the top of container. Plants should be watered with cool, not hot water from a hose at moderate pressure. Keep containers together to increase humidity and water retention. If plants are watered in the morning, they will be dry by the evening and help prevent diseases. Containers placed near reflective surfaces warm up rapidly and lose more water than plants on black surfaces such as blacktop. Containers placed on walks, drives and concrete patios will require more water. Hence, green agro-shed nets can be used to cover vegetables grown in containers during hot summer. Containers are watered daily and so many added plant nutrients are removed from the container. Therefore, they should routinely receive organic materials as fertilizers. Long-season crops like tomato, cucumber, eggplant and pepper may need to be lightly fertilized every 2 weeks or so, to produce a continuous harvest. Tall growing vegetables should be planted on the north side of the garden so that they will not shade low-growing vegetables.

Table 3.1: Planting Information for Growing Vegetables in Containers

Crop	No. of Days From Seeding to Germination	Number of Weeks to Transplanting	General Size of Container	Amount of Light* Required	Days From Seeding to Harvest
Beans	5-8	-	Medium	Sun	45-65
Cucumber	5-8	3-4	Large	Sun	50-70
Eggplant	8-12	6-8	Large	Sun	90-120
Lettuce leaf	6-8	3-4	Medium	Partial shade	45-60
Onions	6-8	6-8	Small	Partial shade	80-100
Parsley	10-12	-	Small	Partial shade	70-90
Pepper	10-14	6-8	Large	Sun	90-120
Radish	4-6	-	Small	Partial shade	20-60
Squash	5-7	3-4	Large	Sun	50-70
Tomato	7-10	5-6	Large	Sun	90-130

*All vegetables grow best in full sunlight, but indicated will also do well in partial shade.

Temporary or permanent containers (including window boxes) can be fitted to any location—balcony, deck, stoops, concrete pad, or any part of the yard. Containers can be located where they are most convenient and where they will grow best (place tomatoes in full sun and the lettuce in partial shade). Better control over growing conditions (water, sunlight, nutrients) can lead to higher yields with less work than a conventional garden. Container gardens are easier to protect plants from weather extremes, insect pests and bigger critters. Vertical growth saves space and allows use of exterior walls.

Table 3.2: Common Problems in Container Gardening

Effect	Cause	Corrective Measures
Plants tall, spindly and unproductive	Insufficient light	Move container to area receiving more light
	Excess nitrogen	Reduce feeding interval
Plants yellowing from bottom, lack vigour, poor colour	Excessive water	Reduce watering intervals; check for good drainage
	Low fertility	Increase fertility level of base solution
Plants wilt although sufficient water present	Poor drainage and aeration	Use mixes containing higher percentage of organic matter; increase number of holes for drainage
Marginal burning or firing of leaves	High salts	Leach container with tap water at regular intervals
Plants stunted in growth; sickly, purplish colour	Low temperature	Relocate container to warmer area
	Low phosphate	Increase phosphate level in base solution
Holes in leaves, leaves distorted in shape	Insects	Use recommended insecticides
Plant leaves with spots; dead dried areas, or powdery or rusty areas	Plant diseases	Remove diseased areas when observed and use recommended fungicide

3.5.4 Advantages of Container Gardening

It's perfect for all kinds of people— people with physical limitation, college students, renters, gardeners, and any gardener wanting to downsize, and save time.

- No digging or tilling is required.
- Container gardening is virtually weed-free.
- It's inexpensive to get started. Few tools are needed.

- Helps to overcome some common gardener complaints like backyards that are too shady for tomatoes, compacted, poor quality soils and soils contaminated with lead and persistent soil-borne disease like *Fusarium* wilt of tomato.

3.6 Hydroponics

It (from the Greek words *hydro* meaning water and *ponos* meaning labor) is a method of growing plants using mineral nutrient solutions; without soil.

3.6.1 Requirements of A Successful Hydroponic System

Avoid big changes in nutrient concentration in the solution because this may damage roots and reduce nutrient uptake. Maintain *p*H in the range of 5-7.5; pH beyond this range would affect availability and uptake of nutrients. Avoid a sharp increase in solution electron conductivity because this can also affect plant ability to absorb nutrients and even damage the roots. Maintain temperature because as the temperature goes up, respiration of plant increases, causing a higher demand of oxygen. At the same time, solubility of the oxygen decreases. Provide a continuous supply of oxygen, as adequate oxygen is the key to hydroponic system.

3.6.2 Setting-up aHydroponic System

x) Container for the nutrient solution: Almost any kind and any shape of the container can be used. The best could be a Styrofoam which holds temperature of the nutrient solution nicely. A container made of wood or bricks lined with plastic can also be used. Plastic sheet for inner lining of the box should be at least 0.15 mm thick to avoid leakage. Depth should be at least 20 cm to provide enough space for oxygen-absorbing roots.

xi) Covering material for the box: This is generally a netting material with a spacing of about 3 mm × 2.5 mm. It will protect plants from insect damage and also keeps rainwater from entering nutrient solution. Also requirement is of some pots or a net bag made from the same net material used for covering and some net for the bottom of the pots.

xii) Nutrient solution: The solution is made up of many basic chemicals which provide both macro as well as micro nutrients. Also needed is some seedling medium like smoked rice hull (The rice hull that has undergone a smoking process). Ordinary rice hull is not effective as a seedling medium. Soil is not recommended either. If smoked rice hull is not available, vermiculite or similar types of seedling media can be used.

Table 3.3: Constituents of the Nutrient Solution**

Element	Chemical Formula	Concentration (ppm)	Amount (g/litre Solution)
N	$Ca(NO_3)_2.4H_2O$	70.0	0.59
	KNO_3	30.0	0.22
P	K_2HPO_4	15.0	0.09
K	KNO_3	38.0	-
	K_2HPO_4	83.8	-
Ca	$Ca(NO_3)_2.4H_2O$	100.0	-
	$CaCl_2.2H_2O$	50.0	0.18
Mg	$MgSO_4.7H_2O$	48.6	0.49

** Using 4N H2SO4 to adjust pH value to 6.0

xiii) Step Involved

- To begin with, fill the box to about ¾ full of the nutrient solution.
- Prepare the pots for planting. Place a piece of netting on the bottom of the pots. This helps prevent seedling medium from coming down and separating root system. It also helps in uptake of oxygen and absorption of nutrient solution.
- Net tray rather than pots can be used to plant large rooted plants such as onion or radish.
- Fill pots about three- fourths with seedling medium.
- Then place the pots into the perforated lid of the box.
- Check that the pots are placed and the solution is 2-3 cm above the bottom of each pot.
- Sow seeds and cover lightly with more smoked rice hull.
- Remember to cover box with plastic netting to prevent insect invasion.
- When raining, cover box with plastic to keep out rainwater.
- Leave plants to grow with little care. As the plants grow roots develop in the box. The roots which are exposed to the air are called O roots and roots which are submerged are called WN roots. The success of the hydroponic system is dependent on the rapid growth and quantity of O roots.

Before too long you can harvest your vegetables; they are healthy, and disease-, insect-, and chemical- free

3.6.3 Basic Requirement for a Hydroponic System

i)   Light- It is essential to carry on photosynthesis. Without this cultivation of vegetables is not possible, irrespective of nutrients provided. Sunlight is the ideal source of light required to produce a healthy growth. However in a hydroponic system, sunlight is not always an option. Artificial horticultural lights can be for hydroponic gardener; these are cost- effective too.

ii)   Oxygen -nutrient ratio: Plants can't absorb nutrients unless oxygen is present. Higher the oxygen level quicker is the absorption. Oxygen maintains a healthy root system and allows plant to absorb nutrients. In a hydroponic system, the water is a medium through which nutrients and oxygen are fed to roots. Make sure to keep roots moist, not soaked.

iii)   Nutrient strength: Nutrients must be solely designed for hydroponics. Soil fertilizers utilize bacteria to break- down complex elements into useful ones; an ideal hydroponic system has minimal bacteria, if any. Soil fertilizers are less soluble; hydroponic systems require solubility, as the nutrient delivery system is based upon this. Soil fertilizers are generally not pH adjusted, and usually too slow to release necessary elements to be suitable for the system.

iv)   Growing media: In hydroponics, growing media, not soil, hold moisture and anchorroots. Composed of inert mineral matter, it won't decompose or harbour potential soil-borne problems. All the plant's nutritional requirements are filled with the nutrient mixes.

v)   pH (alkalinity and acidity): It is the level of acidity or alkalinity of the nutrient solution. Most nutrients in common tap water will be within the range of 6 to 6.5 pH, which is suitable for hydroponic system.

vi)   Temperature: In a hydroponic system, temperature requirement is same as out of a hydroponic system

vii)   Air: Plants require carbon dioxide, it is what they breathe. Poor ventilation would kill plants as surely as a lack of sunlight or water would. Ventilation systems as well as carbon dioxide enrichment and control systems are affordable and available.

viii)   Water quality: In most situations, tap water is just fine for the hydroponic system; over extended periods of time you may get some mineral build-up, but this is not a major cause for concern. Excessive salinity or high zinc content could be harmful to the hydroponic garden.

3.6.4 Benefits of Hydroponic System

Anything can be grown and there is no back-breaking work, and no tilling, raking or hoeing.

No weeds to pull and no poisonous pesticides to spray. No moles or cutworms eat your roots, and most insects leave your clean and healthy plants alone. Ideal for the hobbyist home-owner or apartment dweller who doesn't have the time or space for full-time soil gardening. A hydroponic system distributes nutrients evenly to each plant; their roots are not to be pushed through heavy, chunky soil to compete for nutrients. Hydroponic plants grow faster, ripen earlier and give up to ten times the yield of soil-grown plants. These clean and pampered plants produce fruits and vegetables of great nutritive value and superior flavour.

3.6.5 Problems with Hydroponic

i) **How much nutrient to pour over the aggregate?** Assuming that the container is waterproof and that the inside bottom of it can't be seen through walls or down through the aggregate, it is very difficult to gauge the level of nutrient solution. Without this, it is quite likely that plants would be killed by either under or overfilling. The only simple solution to this problem is to use a see-through container, a transparent inspection window or a float system that will allow a visual check of the nutrient level.

ii) **How often to pour nutrient over the aggregate?** The nutrient solution will evaporate from loose aggregates much more quickly than water from soil. Generally speaking, one may have to supply nutrient to plants about once a day. And one to four times a day may be necessary depending on light, temperature, humidity, what is being grown, how large are plants and size of the container. There can be problem of proper aeration of roots. One of the major reasons for using hydroponic (Fig. 3.8) aggregate is for aeration.

Fig. 3.8

3.6.6 Tips for the Care of the Hydroponics

i) **Cleaning system:** Remove dead leaves before they start rotting; for saving from fungus infection. Keep a close watch for red spider and white fly infestations. They are the two greatest insect problems in hydroponics. Once a year, or after every crop, clean out the system and sterilize growing media. This can be done by picking out the worst of the bits and pieces and then placing the medium in the oven for about an hour at 232-260°C. The system should be flushed every thirty days to remove accumulated mineral hardness left by water additions as the accumulated minerals and salts would slow down plant growth. Flushing is done with plain water. If the system has drain holes, plug them temporarily and fill the planter to the brim. Don't worry about the plants. Let the water stand for about an hour and then drain away. If you are flushing the system because of a nutrient over supply, operate the planter on plain water for a week and only then pour nutrient solution again. .

ii) **Record keeping:** Keep log record of everything done on day to day basis for each planter. Record pH, list of nutrients added, and amount of light, when seeded, when transplanted, first fruit, first harvest and harvest. Anything that has been added later is worth putting into the log. This is especially true when problem is encountered, the log would provide background information.

iii) **Pruning:** When growing indoors, do not allow the top of the plant to get too far from root system. Almost every plant will grow larger in hydroponics than in soil, because plants are getting a full measure of nutrients, air and water. Tomatoes should be pinched off from the tops when they reach a height of 75 cm.

Cucumbers, on the other hand, should be pinched off after seven sets of leaves. Pinching off in this way makes plants manageable under the light and keeps their energy requirement close to root system.

3.7 Urban Gardens

This gardening is the practice of cultivating, processing and distributing food in or around (peri-urban) a village, town or city. The key objectives are to make best use of limited land space, limited water availability as well as recycling of food in the process of sustainable development. Urban gardening contributes to food security and food safety as it increases amount of food available to people living in cities and allows them have fresh vegetables.

Advantages of Urban Garden

It is a sustainable food-based approach and if designed and implemented adequately, has high success rate among agricultural interventions. This is a pro-poor technology which links agriculture with nutrition, health and income. It serves as a mechanism to promote and increase use of nutrient-rich vegetables, particularly indigenous vegetables. It also helps in re-cycling material besides providing aesthetic value to households.

3.8 Roof-top Garden

Building **this** is not easy, but with research, good planning and determination, such a project could substantially upgrade lifestyle of people. A rooftop garden is the ideal alternative way to enjoy all the virtues of gardening and outdoor space when there is no land available. It is ideal for urban environments, where ground space is limited.

- Before we begin, we need to find out if it is possible to create a garden on the roof and the roof is able to hold the weight of a rooftop garden as well as is not damaging the house. Remember to use lighter containers, soil, flooring, furniture etc.
- Work out a comfortable access way to roof.
- Choose a design, figure out how you will layout the roof garden.
- Workout a planting scheme, with plants to work well with natural light you have, humidity, wind etc.
- Get containers/planters, furniture etc. to furnish it. All should be light weight and stable to fix and integrate in the plan/scheme.
- How to water the garden-installing water storage or an automatic irrigation system or by hand depending upon the access to water.etc. Windbreaks, as a rooftop garden will be very windy; they should be completely solid, since they will blow over more easily.
- Heat, wind and heavy rain are the enemies when people want to grow on rooftops. Small platforms will elevate plants slightly above the actual rooftop to increase ventilation around the potted plants.
- Precautions: It is important not to do anything that might damage the property.

3.9 Cultivation Structures for Optimizing Land Space

i) Tower

A vertical structure that can be used to maximize limited land space. Option can also be of an elongated, cement tank to utilize vertical space. Different crops can be cultivated in a vertical mixed cropping system. It is suitable for mainly growing: eggplant, chilli, capsicum, radish, cabbage, tomato, mint, coriander, spinach, basella, kangkong

ii) Frame

It is simple vertical structure used to grow shade -loving plants above ground level. It can be prepared by using coconut branches or wood pieces and digging them into the ground to form a circular, oval or box appearance. It can also be fitted onto a pot to make it portable. It is suitable for growing all vegetables.

iii) Pyramid

It is a vertical structure, but a combination of tower and frame. Number of vegetables to be grown can be increased by altering width and length of the structure and also increasing number of floors. Preferable crops can be as follows:

- Ground floor: vine crops to use aerial space
- Middle floors: leafy vegetables, yams, radish, carrot
- Top most floor: tomato, chilli, eggplant

iv) Wall

It is also made use of to cultivate vegetables. The best location is east-west direction wall , which receives long hours of sunlight. Two options are built –in- pots or fixed- on -pots. The crops suitable are: tomato, chilli, capsicum, leafy vegetables. Vine crops can also be grown if provided with poles or support threads.

v) Bangle

It is similar to cultivation tower. Use re-usable, abounded pots, containers and other materials for growing vegetables. Support is provided by rods at different levels to provide growing space. The preferable crops are leafy vegetables, shade -loving vegetables, tomato, chilli

vi) Mat

It can be used horizontally or at an angle, and is most suitable for creepers or vine crops. A net is weaved or tied to erected poles on which crops are trained. Preferable crops are gourds, beans and yams.

vii) Umbrella

It uses vertical space; a pole with a wire umbrella shaped net cover on the top. Edges can also be used to hang pots; is most suitable for creepers or vine crops

viii) Consortium

It is a combination of several cultivation structures to use vertical space efficiently. It maximizes sunlight and shade for a home garden, and is most suitable for all sorts of vegetables.

ix) Envelope

It **needs** sufficient horizontal land area and is most suitable for water scarcity areas. It makes use of a shallow pit on which a perforated polythene sheet is spread. Shallow pit conserves water and retains it long for growing crop. It is most suitable for leafy and ground cover vegetables.

x) Tat

It is a vertical structure that can be used to maximize limited land space. PVC gutters are used and connected together in series one above the other; suitable for mainly leafy vegetables

xi) Ladder

It is a useful tall structure requiring less floor space and more vertical space; A-shaped frame made from using wood or iron bars with holders to place PVC containers resembling a step ladder. It is most suitable for leafy vegetable

xii) Rack and Trolley

It is shorter than a cultivation ladder, but wider and with a wide angle. It can be made into single-sided or double-sided cultivation. Materials used can vary from PVC pipes to tree trunks, iron or wood; most suitable for growing leafy vegetables.

xiii) Net and Fence

It uses limited horizontal space and any convenient height, and is best suitable for small balconies. In addition to providing fresh vegetables, also makes the landscape attractive; most suitable for creeper vegetables.

xiv) Tripod

It is similar to cultivation ladder but requires even less land space. Height can be increased to adjust many layers and any sorts of vegetables can be grown.

xv) Bags and Portable Towers

They are traditional simple structures which promote recycling of waste materials. It can be either hanging type or stationary type, and can grow any sorts of vegetables.

xvi) Bottles

This aids recycling of soft drink and water bottles. It requires little vertical space and can be hung on roofs, balcony railings, wires, tree branches. Preferable crops are mint, coriander, spinach, kangkong.

4

Art of Developing Home Gardens

4.1 Requirements for Successful Home/Vegetable Gardens

To be a successful home/vegetable gardener, one should have a good knowledge of varieties one grows and of the environmental factors. First and foremost you need to decide which vegetables you want to grow in the garden. Different vegetables have different sunlight requirements — vegetables with fruits require direct sunlight, and leafy vegetables can be grown in a partial shade.

4.1.1 Variety Selection

The selected varieties should be reliable, adapted, and productive, and be of desirable quality and as far as possible resistant to major diseases and insect-pests. They may be open pollinated (seeds produced by natural pollination through successive generations; save seeds from best plants); or hybrids (seeds produced by controlled pollination of two distinct parents; for hybrids buy fresh seeds when needed)

4.1.2 Seeds

The seed can be a cutting, rhizome, bulb or tuber. Seed should be damage- free and free from mixtures of other varieties, from seed-borne diseases, and should have good vigour and germination capacity.

Soil Preparation: For an ideal vegetable garden, soil should have high organic content, be well drained and have high moisture retentivity. Make sure to avoid ploughing soil when it is wet. Remember that different vegetables require different degrees of soil acidity. Hence it is better to have the soil acidity tested before planting vegetables.

To avoid chances of thinning or failed germination make sure that extra seeds are planted in each row. The seeds should be kept moist till they germinate. In case you want to thin your plants, it is better to do so after emergence of 2nd or 3rd set of true leaves.

Use of chemical fertilizers would primarily depend on the natural fertility of the garden soil, amount of organic matter, crop and type of fertilizer. Before using any kind of fertilizers in the vegetable garden, it is advisable to get the soil tested.

To avoid pest infestation in the vegetable garden make sure that plenty of compost is used and there is proper water drainage. Regular crop rotation is beneficial in avoiding pest infestation.

For healthy growth of vegetables, they need to be transplanted at the appropriate time. Transplantation should be according to their temperature requirement and hardiness attained.

For a successful vegetable gardening, weed growth needs to be controlled. Make sure toget rid of weeds that have emerged after rains and irrigations. Also, for a healthy and successful vegetable gardening, water needs to be given properly. During growing season, a vegetable garden requires moisture equivalent to 2.5 cm of rain water; so water your garden thoroughly, at least once in a week.

4.1.3 Land Preparation

If the garden has to be established, soil should either be sterilized through solarization or chemicals to eliminate weeds and soil-borne pathogens. Deep ploughing for loosening of a thick layer of soil improves internal drainage. Harrowing breaks up large soil clods. Add well decomposed FYM. Make proper tilth of soil and level it. Soil should be friable, free from crust formation and disease organisms. Raised beds are preferred (20 cm height). Rows should be with proper crop planting distance.

4.1.4 Sowing Plan

The following points are important: Varieties to be grown; Crop rotation; How much space each crop will use? When? ; Crop arrangement/crop selection; Perennials, height, time in garden; Time of planting; Cropping season; and Successions, inter plantings, relay plantings

4.2 Sowing Methods

4.2.1 Preparing Seeds for Sowing: Soaking or Pre-Germination

This is a common practice for large-seeded crops with hard seed -coats, such as bitter gourd and *Luffa* gourd. Before sowing, seeds are soaked in water and wrapped in damp cloth until germination starts.

*Advantage:*Advantage: It takes less time for the seedling to emerge. This minimizes need for constant watering during dry season and use of labour and irrigation water. The practice can also be followed with small-seeded crops, which are direct seeded. Small germinated seeds are much more difficult to handle as they are prone to mechanical damage. To minimize damage and to facilitate handling, seeds are suspended in a fluid medium and mixture is sown. The process is called fluid-drilling. The fluid medium can be mixed with pesticides, growth hormones and fertilizers to stimulate early growth of the seedlings. Similarly, seeds of tomato, chilli, brinjal which are refrigerated for a long time needs to be conditioned for certain period of time to enhance germination. Seeds can be conditioned by keeping stored seeds in humid place for about 5-6 hours.

4.2.2 Seed Treatment

It is referred to procedures aiming for disinfection of seeds or protecting them against pests, which may pose hazards during germination and in subsequent stages of plant growth. It may be **physical** or **chemical**.

i) Physical treatment may consist of soaking in warm water or applying dry heat. For example, cabbage seeds are soaked in hot water at 45 °C for 20 minutes to control black- rot. In pepper, heating seed in an oven at 76 °C for three days, following a waiting period of three months after harvest eliminates all seed-borne viruses. Heat treatment is not normally a good practice because it tends to reduce germination. The viability of heat-treated seeds decreases with continued storage after treatment. If this treatment is to be used at all, it should be just before sowing.

ii) Chemical treatment usually consists of a use of fungicide, insecticide, or a mixture of both. The chemical can be applied as powder, spray or slurry (thick paste of powdered pesticide and water) at very low rate, approximately 1-5 g /kg of seed. Many commercial seeds are pre-treated before they are sold. The most common fungicides used for seed treatment are Thiram and Captan. Both are broad spectrum ones in action and have low mammalian toxicity. Some systemic fungicides such as Ridomil (metalaxyl) provide protection against fungal disease up to maturity of the plant. Detergents such as sodium or calcium forms of hypochlorite and trisodium phosphate are effective in eliminating seed-borne viruses, particularly carried in the seed- coat. For example, immersion of pepper seed in 100 g/litre trisodium phosphate results in near-complete inactivation of capsicum mosaic virus without affecting germination.

Diluted concentrations of acids such as sulphuric acid are also used for treating seeds against bacteria, fungi and viruses. Among the insecticides, the common materials are Gardona and Malathion, which are very effective against weevils.

Chemical treatment, like heat treatment, may reduce germination. The risk, however, is lesser than heat treatment.

There are certain techniques which are for enhanced seed germination, better plant growth and tolerance to abiotic and biotic stresses — scarification, vernalization, seed hardening, seed conditioning and seed bio priming.

iii) Scarification: This is done on hard seeds such as okra and some legumes. The principle is to soften seed- coat to make a wound on it so that water can be easily absorbed by seeds, thus hastening germination. It can be done by chemical means such as treating winged -bean seeds with concentrated sulphuric acid or physical means such as passing okra seed through a metal brush in a rotating metal drum. After chemical treatment, thorough rinsing with careful timing is required to prevent damage to inner seed structures. Chemical scarification should be done immediately before sowing, otherwise, germination percentage may decrease.

iv) Vernalization: It is a process of exposing germinating seed or plant to low temperature (0-5°C) for a certain period of time to induce early flowering and higher seed yield. Seed vernalization is used for seed production of Brassicas .which are known to respond to this. Radish seeds, for example, are vernalized at 5°C for eight days. Seeds should be immediately sown after vernalization. Vernalized seeds cannot be dried and stored as these lose viability. When aim is to produce fresh vegetables, seed vernalization should not be done.

Among vegetables, radish, Chinese cabbage and mustard are known to respond to seed vernalization; they also respond to plant vernalization. However, cabbage, cauliflower and in favourable conditions, these tend to flower more. The effect of vernalization may be partially or fully reversed by high temperature in production field. The phenomenon is sometimes called de-vernalization. It explains seemingly lack of response to vernalization by plants that are grown under relatively high temperatures.

v) Seed Hardening: Like vernalization, seed hardening is a treatment that is applied to germinating seeds, and its effect is seen on the developing plant. The process consists of air-drying of seeds, which have started germination but have not produced any radicle. The hardened seeds are then sown immediately. Hardening makes seed emergence faster and more uniform. It also promotes faster seedling growth, better root-shoot ratio, and better transplant survival. The treatment is best done on slow-germinating seeds such as tomato, eggplant and pepper. Hardening can also be done at the seedling stage for all transplanted crops. In this case, seedlings are allowed to wilt in seed- bed by reducing frequency of watering and exposing seedlings to full sun if they were previously grown under partial shade. Hardening process is started ten days before transplanting. Its results are better root-shoot ratio and transplant survival.

vi) Seed Conditioning: In some cases, seeds may have been over- dried and stored at low moisture content (such as canned seed stored at low temperature); such seeds do not easily absorb water and exhibit poor germination. This can be corrected by exposing seeds to high humidity for one to two days before sowing. A practical way is to put seeds on a wire screen tray suspended in a sealed jar with water, without wetting seeds. The procedure improves germination of seeds such as pepper.

vii) Seed Biopriming: It is a new technique of seed treatment that integrates biological (inoculation of seed with beneficial organisms to protect seeds) and physiological aspects (seed hydration) of disease control. Recently, It has been used as an alternative method for controlling many seed- and soil-borne pathogens. Seeds are soaked in liquid bioformulation and later coated with dry bioformulation. It is an ecological approach using selected fungal and bacterial antagonists against soil- and seed-borne pathogens. Biological seed treatments may give an alternative to chemical control.

4.3 Garden Tools and Jobs they Perform

a) Compressed-air Sprayer

It is the most popular piece of equipment for applying pesticides,ait gives good coverage, especially to underside of plant leaves.

b) Hand Cultivator

It is for working around vegetable plants and breaking- up soil clods for light replanting.

c) Hand Duster

It is used to apply pesticides in powder form.

d) Hand Seeder

Some seeders can open furrows, drop seeds, and also cover seeds in one operation. Majority of the hand seeders are adapted to a wide variety of seed sizes.

e) Hoe

It comes in all shapes, sizes and models. It is used for preparing seedbed, making rows and for cultivating soil to mix in fertilizer and control weeds.

i)   A common or square-bladed- filed hoe is good for most garden jobs.

ii) A pointed or Warren- hoe is good for opening a furrow by string and for cultivating between plants.

iii) A scuffle- hoe, made in several patterns, with a flat bottom that cuts weeds off under the soil surface and breaks- up the crust layer on top of the soil as it is pushed back and forth between rows.

f) Irrigation Equipment

It is a watering can for transplanting; with garden and soaker hoses and sprinklers for general watering.

g) Long-handled Cultivator

It breaks up large clods and refines seedbed.

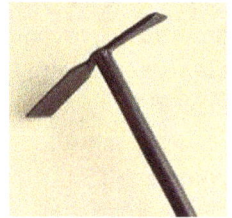

h) Measuring Stick

It is for determining distance between plants and rows.

i) Rake (for Leveling and Grading Soil, Stirring up Soil Surface, and Removing Clumps, Rocks, and Shallow-rooted Weeds)

i) A bow rake is good for smoothing out soil, removing stones, and breaking up clods.

ii) A straight rake is designed so that its back can be used to smoothen seedbed and to compact soil over freshly sown seed for improved germination.

j) Secateurs

It is used to cut various types of branches and unwanted growth.

k) Spade (for Edging Beds, Digging Holes for Planting, Slicing Under Turf and Working Soil Improvements into the Garden)

i) A spade with a sharp edge is used for turning soil and incorporating organic matter.

ii) A four-pronged fork is good for mixing compost pile.

iii) A round-pointed shovel is good for turning soil and can also be used to harvest large-rooted crops such as Irish potatoes and sweet- potatoes.

l) Spreaders

It is for lime and fertilizer application.

- A drop spreader covers less area than a broadcast spreader with each
- pass over the site, but the area covered is easier to detect.
- A broadcast spreader applies materials uniformly, although margins
- of the area covered may be difficult to mark.

m) String and Stakes

It is for row alignment.

n) Tiller

It makes soil preparation easy for serious gardeners who will use it enough to make the purchase worthwhile. Three types are available; all of them are driven by gasoline or electric motors. One of the most common and least expensive types has tines mounted in front. A second type has the tines mounted in rear. Although more expensive, the rear-tine type is easier to operate. Many tillers with rear- mounted tines have a reverse gear that makes it possible to work in cramped areas. A third type is the center-mounted or mid-tine tiller, which combines advantages of both the other two types.

o) Trowel

It is used for transplanting vegetable plants.

p) Wheel Cultivator

It is for removing weeds and preparing soil.

q) Wheelbarrow or Garden Cart

It makes much easier moving of soil, stones, tools and harvested vegetables.

4.4 Planning the Garden: Cropping Systems and Role of Vegetables

The most suitable way is to plan according to the cropping systems: cool season and warm season vegetables. A good option to reduce plant pest problems is to alternate cool and warm season areas of the home garden each year.

a) Cool Season Vegetables

They require cool weather to grow and mature properly. These can withstand frost, are planted in the early spring and again in autumn. These include lettuce (10-20°C), pea (13-18°C), radish, cabbage (15-20°C), onion (13-24°C) carrot (16 °C), potato and spinach (16-18°C). Also temperate/cool season vegetables are celery (16-21°C), tomato (21-24), Frenchbean (16-30°C).

b) Warm Season Vegetables

They require warm weather to grow properly and are planted after soil has warmed up. Many warm-season crops also need a long- growing season e.g. Cucumber (18-30°C), pepper (21-24°C).

4.4.1 Methods of Planting Vegetable Crops

Vegetables can be classified into three categories, depending on the planting practice: (i)Crops that are usually transplanted; (ii)Crops that are usually direct-seeded, and (iii) Crops that should be direct-seeded

- Crops usually transplanted are cabbage, Chinese cabbage, broccoli, cauliflower, tomato, eggplant, pepper, bittergourd, onion, celery, lettuce, etc.
- Crops usually direct-seeded are watermelon, squash, cucumber, bittergourd, onion, beans, cowpea, soybean, pak-choi, kangkong, yard long bean, etc.
- Crops which should be direct-seeded are radish, turnips, carrots, beets, etc.

Method of Planting Depends on the Following Factors

a) Cost and Availability of Seeds

Direct-seeding always requires three to four times more seeds than transplanting. When cost of seed is high, as in hybrid seed, transplanting may be recommended.

For example, hybrid bittergourds are recommended for transplanting; but open-pollinated (OP) varieties of the same crop are always direct- seeded. A clear exception to this rule is the case of hybrid Granex of onion which is direct-seeded as it is able to produce bulbs of good size even in a relatively dense plant population. On the contrary, open-pollinated Red Creole is transplanted because this variety tends to produce smaller bulbs when grown at high densities. Transplanting of onion gives better control of plant population and avoids densely populated spots.

b) Quality of Land Preparation

Direct-seeding of small-seeded crops is impractical when the field is not thoroughly pulverized during land preparation. Large soil clumps make it difficult to control depth of seeding, resulting in poor emergence. As a rule, small-seeded crops, such as lettuce and celery, should not be direct seeded. Also, when land preparation is inadequate, weeds can be a serious problem in direct-seeded, slow-growing crops, such as onion and celery.

c) Root -regenerating Ability of the Crop

Some crops such as legumes, do not easily regenerate roots, hence, do not easily recover from transplanting shock. In contrast, solanaceous crops and crucifers easily regenerate.Cucurbits are intermediate in rooting behavior, and can be successfully transplanted if the procedure is done early enough at the cotyledonary leaf stage.

d) Multiple Cropping

When vegetables are grown after another crop, it is often advisable to start seedlings even before the crop is harvested. This allows planting of the seedlings immediately after harvesting the previous crop, reducing the period of planting when the field is ready.

e) Rapid Growth Rate

Some vegetables such as sweet- corn, cucumber, and yardlong bean germinate quickly and grow fast. They are easily established in the field even when conditions are not ideal. Hence, they are usually direct-seeded, unlike pepper and celery which grow very slowly in the initial growth stage.

Overall, if the conditions are favourable, it is practical to transplant. This practice allows intensive management of seedlings in a small area of the seedbed in the open field. The result is a good start for the crop, which often is translated into higher yields and better product quality.

4.5 Direct Seeding

It establishes vegetables by sowing seeds directly onto the site, where it is to be grown. It also involves transplantation of asexually propagated materials e.g. bunching onions, potato and sweet -potato.

4.5.1 Vegetables Suitable for Direct Seeding

Colocasia (roots), beetroot, bittergourd, carrot, coriander, cowpea, cucumber, Frenchbean, amaranth, jute, kohlrabi, onion, peas, pointed-gourd, potato (tuber), radish, red amaranth, soybean, sweet- corn, sweet- potato (cuttings), pumpkin, okra, sword bean, turnip.

Good site preparation and effective weed control are essential for success. Direct seeding can be done by drilling or broadcasting. "Drill" is a very small furrow (narrow groove) made on the top of the bed for planting seeds in a row.

For **broadcasting,** first the top of the bed is smoothened. Then seeds are scattered/ broadcast and firmed into the soil with the back of a hoe. Once plants grow, beds can be thinned several times. Farmers/Gardeners rarely use broadcast method for planting vegetable crops, except the following: kale, non-heading Chinese cabbage (pak-choi), peas, beet, carrot, coriander, amaranth, lettuce, water convolvulus (kangkong), non-heading leaf mustard, radish and turnip. These crops require close spacing, mature early (less than 50 days), have cheap seeds, and grow relatively fast.

The **broadcast method** is feasible when the field/plot is adequately prepared i.e., well-pulverized, weed- less and irrigated by sprinklers. Beds approximately 1.0 – 1.5 m wide are used in broadcast method. The most important factor in direct -seeding is planting depth. A consistent planting depth is possible only in a well-prepared plot.

Soil particles should be fine when small seeds such as of carrots and non-heading Chinese cabbage are planted. For large- seeded crops like watermelon, need for thorough land preparation is less. Seeds should be placed deeper in light (sandy) soils to save them from dessicating. Shallow planting is required in heavy soils. The soil cover after setting should be about five times the diameter of seeds. The soil should be irrigated immediately after seed sowing to create favourable condition for germination. It should be kept moist until seedlings are established; then water can be applied less frequently.

4.5.2 Why Direct Seed?

Some seeds germinate rapidly and their seedlings grow fast. Some seeds are large, thus, can be planted in a wide range of soil conditions. Some vegetables like carrot and radish have only one long tap -root system which, if damaged, will deform root. Some have slow root regeneration capacity

a) Advantages

- Direct seeding is much cheaper and requires minimal labour.

- Higher plant density after germination provides better shelter to new seedlings and reduces weed competition. It also allows natural selection to sort out stronger from weaker plants without creating gaps to be replanted.

- The plants are usually healthier and have stronger, deeper root systems as they are not transplanted and there is no disturbance to root growth. This enables plants to be tolerant to stressful conditions such as pest attack and drought.

b) Disadvantages

- Direct seeding is limited to plants, which grow readily from seeds.

- A large amount of seeds is required. Hence, if only minimal seeds are available for a particular species, it is better to raise seedlings for that species in a nursery.

- Plants germinating under field conditions are extremely vulnerable. Frosts, spring droughts, or flooding of the sown area can dramatically reduce seedling establishment.

- The initial density of plants is harder to control.

4.5.3 Transplanting

Vegetables commonly set into home garden as transplants include: tomato, pepper, broccoli, brussels sprouts, cabbage, cauliflower, eggplant, lettuce, tomato and pepper. Other vegetables sometimes set as transplants include kohlrabi, watermelon, summer squash, okra and cucumber.

Advantages of Using Home Transplants: Transplants allow to replace early-harvested vegetables immediately and to produce another crop quickly. Yield losses from poor germination are eliminated. There is control of factors such as cultivar, plant size, container material and its size. The chances of introduction of insects or diseases into the garden are reduced.

4.5.4 Vegetable Nursery

The nursery for growing transplants may be as simple as a raised bed in selected corner of the field (usually near the water source), or as sophisticated as a glasshouse with micro-sprinklers and an automatic temperature control system. Nurseries provide the following conditions for growing seedlings.

a) Protection from pests, including higher animals: In a simple nursery, improvized fences such as bamboo or nets are provided. In screen- houses, protection is better since seedlings are totally enclosed. Fine mesh enclosures provide protection against many insect- pests.

b) Protection from rain and sun: Excessive rain causes waterlogging in the seedbed, which may result in physiological damage to seedlings. Excess moisture favours development of diseases. The only way to control rain damage fully is to provide transparent roofing, but it is costly. Rain damage is also controlled simply by providing nets, mulches or equivalent devices to reduce size of raindrops that fall on the seedbed. Adequate drainage is provided by raising seedbeds.

Excessive sunlight may also cause damage to newly germinated seedlings. To protect them, partial shade such as coconut leaves or nets are used. However, the shade should be removed as soon as the seedlings are established. Prolonged shading may result in spindly and weak seedlings; many of which may eventually die.

c) Protection against temperature extremes: In tropics (except in elevations exceeding 2,000 m where frost may occur during some parts of the year), the ambient temperature is normally suitable for seedling production. However, when seedlings are grown in glass-or plastic-roofed greenhouses to protect them from rain damage, they may suffer from excessively high temperature during sunny days. The damage can be direct (physiological damage) or indirect (by favouring development of diseases and multiplication of insect- pests). Thus, a good tropical glasshouse must provide means for controlling temperature build-up. iv. Germination and seedling growth medium: Soil is the universally available medium for germinating seeds and growing seedlings. However, it is not necessarily the best medium. Some soils are unsuitable for growing seedlings. Special mixtures of perlite, vermiculite, and peat are commercially available in ready-to use mixes for specific purposes, and are used as substitute for soil. The tropical environment is, in fact, rich with materials that can be utilized in formulating nursery mixes.

To help vegetable gardeners to choose materials for raising seedlings, the following characteristics of an ideal nursery medium should be considered.

Fig. 4.1: Materials for Ideal Nursery Growing Medium

d) Seedling growing media: Water-holding capacity and aeration: Organic materials such as peat, are able to retain moisture without causing waterlogging. This characteristic is crucial because germinating seeds and roots of seedlings need both water and air. In contrast, sandy soil tends to lose moisture very quickly, and clay soil tends to retain too much moisture so long that air supply in the root zone becomes restricted. Coconut coir dust, rice husk, mosses, and dried (fully decomposed) manure are also used in a nursery mix to improve soil's aeration and water-holding capacity.

e) Capacity to supply plant nutrients: Vermiculite, perlite, and their tropical counterparts such as coconut coir-dust and rice- husk, are essentially inert and contribute very little, if at all, to plant nutrition. So, mixes that are predominantly made up of these materials need elaborate fertilizer supplementation. Thus nutrient-rich materials such as compost, manureand fertile soil, should be added to this nursery mix. The *p*H of the soil mix should be adjusted to a range of 6-7 to ensure nutrient availability.

f) Freedom from soil-borne plant pathogens: The soil contains millions of different kinds of microorganisms; many of these are helpful to plant, while some are diseases causing. The nursery mix should retain only beneficial types. To achieve this, the nursery mix is often sterilized, either with the use of heat or chemicals.

4.6 Method of Raising Seedlings

4.6.1 Seedbed Method (Fig. 4.2)

Seedlings are raised in beds when large quantities of seedlings are needed as in the community gardenPrepare the bed and improve the soil condition. Sterilize beds either by burning straw on the soil surface or by pouring boiling water on beds. Sow seeds and cover with straw. Water bed using fine sprinkler. Allow seedlings to grow 5-cm apart, thin out excess seedlings when first true leaves appear. Use simple structure to protect seedlings against rain and sun.

Fig. 4.2: Seedbed Method

Disadvantages: Pulling seedlings during transplanting causes a lot of damage to roots. And pulling seedlings with a ball of soil can be laborious and may cause transport problems. Also spread of diseases within the seedbed is difficult to control.

4.6.2 Seed Box or Tray Method/Seed Flats (Fig. 4.3)

Seedlings can also be raised in specially made wooden boxes or plastic trays. It is essentially a portable seedbed. The growing medium consists of soil, sand, compost and smoked rice hulls. The medium is sterilized as in the seedbed method. Sow and cover seeds and then water using fine sprinkler or watering can. Thin or prick out excess seedlings when first true leaves appear. Seed flats are designed to be carried to field where they are emptied of seedlings. Well-designed seed flats are light and sturdy; they also allow extraction of seedlings with minimal root damage. Almost all modern seed flats have cellular designs.

Fig 4.3: Seed Box (Tray) Method

Advantage: Its advantage over the seedbed is that it can be set on tables with slatted tops (wire mesh or bamboo). This exposes seed flat to light and air and prevents outgrowth of roots from drainage hole. Well-branched roots that are not damaged during transplanting, recover easily from" transplanting shock"

4.6.3 Seedling Container Method (Fig. 4.4)

Raising seedlings in separate pots or containers gives 100% survival since root injury is minimized. Seed containers may be small plastic bags or made of biodegradable materials such as rolled banana leaves, paper pots, straw etc. The growing medium consists of soil, sand, compost and smoked rice -hulls. Sow and cover the seeds. Water and thin out excess seedlings when first true leaves appear.

Fig. 4.4: Seedling Container Method

4.7 Sowing and Pricking-out

Sowing can be done using the following three methods.

- **Broadcasting:** Its least labour-intensive but seedlings are not evenly distributed; some of them are too crowded.

- **Drill sowing with uniform spacing:** It can also be mechanized.

- **Sowing at high density in nursery bed, then transplanting (pricking-out) newly emerged seedlings into another seedbed, with uniform spacing:** Its most labour-intensive as far as the planting/transplanting operations are concerned; however, it may save labour in watering a bigger nursery bed during germination phase.

4.7.1 Covering the Nursery Bed☒

Fig. 4.5: Nursery Bed and Nursery Tray Cover with Nylon Mesh Net

Nursery seedling should be covered with nylone mesh net (40-60 mesh size) after germination and after removing straw. This would help preventing insect-pest infestation at early stage and thus prevent insect damage and virus disease spread.

4.7.2 Watering Seed Bed

Watering seedbed should be done very carefully until seedlings emerge, especially when the seeds are small. Large water drops tend to erode thin soil covering of small seeds and seeds may dry up. Watering with a mist sprayer is recommended for highly delicate seeds such as lettuce and celery. As a rule, seedbed should be kept moist but not wet until germination. Mulching seedbed immediately after sowing helps prevent erosion of soil cover and conserves moisture. Watering during seedling production should be done preferably in the morning. If watering needs to be repeated, this should be in the early afternoon. Watering in late afternoon causes surface of seedbed to remain moist at night, a condition favourable for development of damping-off disease. Ten days before transplanting, watering should be reduced to allow shoots growing slower and roots growth faster. The seedlings should also be exposed fully to sun if they have been kept under partial shade. This process is called seedling hardening and assures high transplant survival and quick recovery after transplanting.

4.7.3 Fertilizing Seed Bed

Fertilizers applied to the seedbed must be mixed with soil before, and not after sowing. This is necessary because seedlings must be able to use fertilizers immediately after it has developed root system for absorbing nutrients. Seedlings stay in the seedbed for only 20-30 days, on an average. Fertilizers in granular form applied late may not be available to the seedlings immediately and may cause temporary nutrient deficiency. In seed flats, constant watering causes rapid loss of soluble nutrients. Only fully decomposed organic matter should be used because decomposing organic matter contains a lot of micro-organisms which may compete with seedlings for nutrients. In some instances, it is also necessary to apply fertilizers as drench (starter solution) or as foliar sprays. They should only be used as corrective measures and should not substitute for sound planning and preparation of nursery soil mix.

4.8 Problems in a Nursery

4.8.1 Damping-off

A seedling disease commonly caused by fungi of genera R*hizoctonia* and *Phythium*, and disease organisms are favoured by warm wet nursery beds.

Symptoms: water-soaked lesions are seen on the stem of seedling at the point of contact with the soil. These lesions soften stem, causing seedling lodging ,which eventually dry up and die. The disease organisms multiply rapidly at night in dark and damp conditions; the disease can be controlled by making night conditions unfavorable for pathogens.

Control: Keep nursery bed relatively dry at night by avoiding watering in late afternoon. Soil sterilization, drenching with Brassicol, proper plant nutrition enabling seedlings to mature more quickly protects seedlings.

4.8.2 Oversized Seedlings

These resulting from delayed field preparation and unfavourable weather conditions are more difficult to handle during transplanting. They have lesser chances of surviving in the field, tend to suffer more injury, and have lesser ability to regenerate roots. Such seedlings can be avoided by prolonged hardening. Some seedlings, such as of onion, can be pruned at the top to reduce transpiring surface and making root-shoot ratio favourable for water balance.

4.8.3 Chemical Toxicity

Ammonia toxicity may occur when soil is sterilized with heat.

Symptoms: Yellowing of leaves is similar to iron deficiency. It is more acute when fresh manure is used as a component of nursery mix. Toxicity may also result from residues of chemicals used in soil sterilization. Chemical sterilants should be carefully selected, and prolonged heat sterilization should be avoided.

4.8.4 Additional Problems

There are additional challenges are as follows:

Trouble Shooting	
Common Problems	**Causes(s)**
Tall, straggly seedlings	Light intensity too low
	Nitrogen fertilization too high
	Night temperature too high
	Plants spaced too close
Older leaves yellow	Nitrogen fertilization needed
Seed doesn't come up	Seed old or improperly stored
	Too wet or too dry
	Temperature too low
	Seed planted too deep
Seedlings look pinched at soil line, fall over and die	Damping off:
	Do not overwater
	Grow at proper temperature
	Use sterile media and containers
	Grow under strong light
Purple leaves	Phosphorus deficiency
	Temperature too low

4.9 Preparing Seedlings for Transplanting

When dry and sunny weather is immediately after transplanting, seedlings need to undergo hardening. In non-cellular seed flats, such as wooden flats and seed beds, it is a good practice to prune roots one week before transplanting by passing a sharp knife around each seedling. This procedure stimulates root branching close to main root and assures good root-shoot ratio. The dense root system immediately around main root serves to hold nursery soil and prevents bare-root transplanting. Transplanting from seed beds or non-cellular seed flats usually causes damage to roots that are too far from tap- root. A starter fertilizer solution of 0.1% urea or ammonium sulphate is applied to seedlings just before transplanting. This assures ready supply of nutrients for recovery of roots.

Preparing Plot for Transplanting

If the field is dry, it must be irrigated thoroughly a few hours before transplanting. The irrigation is preferably localized along the plant rows, leaving areas between rows dry for transplanting. Watering immediately before transplanting goes with light sandy soil but not with clay soils. Clay soils tend to be sticky and difficult to manage during transplanting. Manure and fertilizer should be applied before transplanting and mulching. If grass mulch is to be applied, this must be spread carefully after transplanting to prevent seedling damage. Plastic mulch should be applied before transplanting.

Importance: Land preparation is for eliminating most weeds and soil-borne pathogenic microorganisms and to improve water- holding capacity, drainage and soil aeration to facilitate field operations such as furrow irrigation and mechanized weed control.

4.10 Steps in Land Preparation

4.10.1 Clearing/Mowing

This is an optional step necessary only in opening new areas for vegetable growing and in preparing plots after a prolonged mismanaged fallow (rest) period. It is necessary to clear plot of obstructions and also of tall weeds.

4.10.2 Tillage

After the plot is free of obstructions, it is pulverized, levelled, and ridges are formed.

Loosening of a thick layer of soil improves internal drainage. If the soil is too hard to achieve maximum digging depth at the first passing, second digging should get the right depth, Success in digging is determined to a large extent by soil moisture, which should be ideal. Digging wet soil may result in the formation of hard soil clods that may not easily be broken. Extremely dry clay soils, on the other hand, may be too hard.

4.10.3 Levelling

A levelled plot with a slight grade allows even distribution of water through the furrow method. Leveling should be done before fertilizer and manuring.

4.10.4 Ridging

The final step in land preparation is making furrows or beds for planting. The size of beds or distance between furrows depends on season, soil type, irrigation method and crop. In heavy solid, wider beds and in lighter soils, narrower beds are advised.

During wet season, single-row of planting is recommended to minimize competition for space between rows of luxuriantly growing plants; consequently, narrower beds are prepared. The reverse is true during dry season when plants tend to be less vegetative. Multiple-row beds tend to reduce irrigation water loss by evaporation; hence they are preferred over single-row beds for dry season crop. Viny crops and those with big canopies need spaces between rows. On the other hand, onions are planted in multiple-row beds with a spacing of 20- cm between rows.

4.10.5 Manuring

Manure can be applied immediately before planting if the material is sufficiently dry and decomposed. Fresh manure tends to generate heat and ammonia during decomposition, and may harm crop. Chemical fertilizers may be broadcast like manure or applied along the crop row during planting. The latter is preferred because it allows efficient utilization of fertilizers

4.10.6 Transplanting Stage

Tomato, pepper and eggplant: The seedlings become ready at 3-6 open true leaves with the stem height of 10-15 cm.

Fig. 4.6: Tomato and Pepper Seedling Ready for Transplanting

Cabbage, chinese cabbage, cauliflower, kohlrabi and lettuce: Their seedlings are ready at 4-5 true leaves stage or 3-5 weeks after seed sowing.

Fig. 4.7: Cabbage, Brinjal and Cauliflower Seedlings Ready for Transplanting

4.10.7 Ideal Conditions for Translanting

A cloudy, cool weather and moist but not wet soil are ideal for transplanting. During sunny days, transplanting is best in the late afternoon to give time for seedling recovery at night. Seedlings that are adequately hardened with slightly damaged roots recover well when transplanted in a well-irrigated field, even on a hot day.

Fig. 4.8: Transplanting at Right Depth with Proper Method

4.10.8 Transplanting Shock

Transplanting shock refers to the temporary growth retardation or mortality of seedlings after transplanting.

Prevention: This is through adequately preparing seedlings as well as plot for transplanting.

Seedlings can recover easily if watered frequently for a week after transplanting. Protection during extremely hot days can be provided by banana bracts. This procedure is generally labour-intensive and is not normally required if seedling preparation and transplanting procedures are carefully followed. It is practical in small-scale farming.

4.11 Cropping Systems

4.11.1 Sequential Cropping

A form of multiple cropping in which crops are grown in sequence on the same plot with succeeding crop planted after preceding crop is harvested. It is the planting of two or more crops, one after another, in the same plot to maximize land productivity. In the low elevations of tropical Asia where rice is the dominant crop, one or two vegetables crops grown between two wet-season rice crops is a common sequential cropping pattern.

4.11.2 Pest and Weed Control

Weeds are controlled effectively when land is used continuously for growing crops. A prolonged fallow or rest period between crops increases weed population, particularly in areas with even rainfall distribution; as this condition favours growth and multiplication of weeds. This problem is lesser in areas extremely dry during fallow period. Proper sequencing of crops, so that two successive crops do not share the same disease or insect problem, can effectively break life cycle of pests; eventually reducing pest population and makecontrol easier.

4.12 Residual Soil Moisture and Nutrients

Fertilizers applied on the previous crop can also be utilized by the succeeding vegetable crop.

The balance of nutrient in the soil can be maintained by rotating crops of different nutrient utilization patterns. Leguminous vegetables can fix atmospheric nitrogen and return some to the soil at the end of the season. When planning a sequential cropping system, crops which can be transplanted should be grown in the seed bed before the harvesting of current crop. This shortens growing period of the succeeding crop. Crops should be carefully selected, taking into account their most favourable planting date. Early- maturing crops are generally preferred to allow growing of more crops per year. Crops belonging to the same family should not be planted in succession to avoid accumulation of pests. For example, eggplant should not follow tomato or pepper because this may build up bacterial wilt; a soil-borne disease that affects all these crops. Radish should not follow cabbage because both are hosts to diamondback moth, an insect- pest difficult to control. Some crops produce root exudates which may remain in the soil and harm next crop. Decomposed residues may also cause damage to next crop. To make full use of soil nutrients from different strata, deep-rooted crops should be grown in rotation with shallow-rooted crops.

4.13 Intercropping

It is the practice of growing two or more crops in close proximity. The most common goal of intercropping is to produce greater yield on a given plot by making use of resources that would otherwise not be utilized by a single crop. Intercropping is the practice of growing two or more crops in the same field at (or about) the same time. In selecting crops for intercropping, it is better to plant companion crops which don't compete.

Table 4.1: Companion and Antagonist Vegetable Crops

Vegetable	Companion Crop	Antagonist Crop
Asparagus	Basil, parsley, tomato	Garlic, onion
Beans	carrot, cauliflower, cabbage, cucumber, potato most other vegetables and herbs	
Bottle gourd	Sponge gourd, cucumber, bitter gourd	Onion
Bush beans	potato, cucumber, corn, Celery	
Pole beans	Corn	Onion, beet, kohlrabi
Beets	Onion, kohlrabi	Pole beans
Brassica crops	potato, celery, tomato, mint, beet, onion	pole bean
Carrot	Peas, leaf lettuce, onion, leek, tomato	
Celery	tomato, bush beans, cauliflower, cabbage	
Corn	okra, tomato, bush beans, pole beans, cabbage, vine squash, potato, peas, cucumber	
Cucumber	Pole beans, radish, okra, eggplant, beans corn, peas, radish	Potato
Eggplant	Beans, kangkong, vine squash, Chinese cabbage, Radish	
Kangkong	Tomato, okra, corn, eggplant, amaranth Any crop on trellis	
Leek	onion, celery, carrot	
Moringa	Kangkong, Chinese cabbage, nightshade, lettuce, bush squash, amaranth	
Lettuce	Carrot, radish, cucumber	
Mungbean	corn	
Okra	kangkong, vine squash, Chinese cabbage, radish,	
Onion, garlic	beet, tomato, lettuce, carrot	Peas, beans

Contd...

Vegetable	Companion Crop	Antagonist Crop
Parsley	Tomato, asparagus	
Peas	carrot, turnip, radish, cucumber, corn, beans,most vegetables and herbs	Onion, garlic
Potato	Beans, corn, cabbage, squash, cucumber,	tomato, Potato
Squash	Corn, peas, lettuce, cucumber	
Radish	grows with anything, helps everything	
Soybean	bottle gourd, cucumber, bitter gourd	
Luffa gourd	onion, lettuce, asparagus, carrot, radish, chinese cabbage, Kangkong, vine squash	
Tomato		Kohlrabi, potato, fennel, cabbage
Turnip	Peas	

Advantages: There is better use of light, nutrients and water by allowing intercrop to utilize these resources that may otherwise be wasted. A crop which covers soil rapidly without competing excessively with the associated crop, can reduce growth of weeds and prevent them from competing with the main crop.

4.13.1 Yield Stability

It is often claimed that major reason for predominance of intercropping in poorly developed home gardens is that it give greater stability of yield over seasons. For example, if cabbage is intercropped with corn and cabbage is attacked by diamondback moth, the performance of corn will not be affected as corn is not a host of diamondback moth. If cabbage is planted as a monocrop, then entire plot can be lost. On the other hand, if corn is attacked by corn borer, the performance of cabbage will not be affected. On the contrary, the loss of the corn crop may create a favourable condition for cabbage, because the shading effect and possible competition of corn with cabbage for nutrients and water would be reduced.

In tomato and cabbage intercropping, the tomato is believed to serve as repellent against diamondback moth. Intercropping serves as an insurance against uncertainties. Selection of species for intercropping must take the following into consideration.

- allelopathy and residue problems;
- depth of rooting;
- combining crops with different nutrient demands;
- combining tall crops with short but have tolerant ones. Tall crops, such as corn and okra, and crops that are grown on stakes such as indeterminate

tomato pole beans, bittergourd, and cucumber can be intercropped with shade-tolerant crops such as celery, Chinese cabbage, green onion, and cabbage; and

- growing short-season crops between late-maturing crops.

This practice allows utilization of space between rows of long-duration crops. For example, fast growing leafy type Chinese cabbage (pak-choi) is transplanted and harvested within 30 days between rows of eggplant, even before the flowering begins in eggplant . Light is considered the most critical factor for intercropping with regard to temporal use of resources. The amount of light intercepted by a crop canopy depends mainly on its leaf area. The plant population and spatial arrangement of associated crops should be planned to reduce interspecific competition and to enhance light penetration. In intercropping, competition for soil nutrients between component crops does not occur until there is an overlapping of depleted resources by associated crops.

4.13.2 Succession Cropping

It is planting two or more different vegetables in sequence in the same garden space within one growing season. Succession cropping permits several plantings of certain well liked vegetables without causing disease build-up. Space between tall height crops can be well utilized. Nutrition as well as water can be properly used, for example growing spinach, coriander, lettuce, amaranthus between chilli, eggplant, tomato, cabbage and cauliflower.

4.13.3 Relay Cropping

It is a form of multiple cropping in which there is growing of two or more crops in the same space during a single growing season. It is a type of polyculture. It can take the form of double-cropping, in which a second crop is planted after the first has been harvested, or relay cropping, in which second crop is starts amidst the first crop. The relay crop must be shade-tolerant; otherwise, the population of the first crop must be reduced to minimize shading effects on the relay crop. Relay crop preferably be drought-tolerant, as it may not be irrigated. The disadvantage of relay-cropping is that it requires more labour.

4.13.4 Companion Cropping

it is the planting of different crops in proximity (in gardening), on the theory that they assist each other in nutrient uptake, pest control, pollination, and other factors necessary for increasing crop productivity. For gardeners, combinations of plants also make for a more varied, attractive vegetable garden, as well as allowing more productive use of space.

Some plants do not grow well with other plants and some do better next to certain types of plants. Some plants secrete growth-suppressing chemicals like beans so do not follow onions;,tomato/potato should not be followed by Brassicas ; and gourds not to be followed by onion and garlic. Some plants have insect repellent properties like onion, garlic, marigold and some plants attract beneficial insects like mustard, lily, marigold.

Following are the companion plant to deter the insect- pests.

- Basil/Osmium   : flies, mosquito, tomato borer
- Garlic   : beetles, aphids, weevils, spider mites, carrot fly
- Radish   : cucumber beetle
- Mint   : cabbage moth
- Marigold   : beetles, cucumber beetles, nematodes
- Nasturtium   : aphid, beetle, squash bug,
- Rosemary   : cabbage moth, been beetle, carrot flies
- Tansy   : beetle and flying insects
- Petunia   : Beetles on beans
- Peppermint   : Whitefly
- Tomato   :cabbage diamondback moth

4.14 Crops Fitting in Home Garden

Crop fitting depends on: Total crops/plots at a time or in a season; Cropping seasons (sowing and harvesting time); Family requirement/preferences; Diversified diets; Nutrition requirement and supplementing crops; Soil nutrition; Cropping system

Table 4.2: Sowing and Harvesting Time of Common Home Garden Vegetables

Crops	Sowing Period	Transplanting Period	Harvesting Time
Amaranthus	Feb.-July	-	April-Oct.
Basella	Feb-Nov	-	Round the year
Beat Root	Oct.-Nov.	-	Dec.-Feb.
Bitter Gourd	Feb.-March June-July	-	May-July Aug.-Oct.
Bottle Gourd	Feb.-March June-July	-	April-June Oct.-Dec.

Contd...

Crops	Sowing Period	Transplanting Period	Harvesting Time
Brinjal	Jan.-Feb.	Feb-Mar.	April-June
	May-June	June-July	Sept.-Nov.
	Oct.-Nov.	Jan.	March-May
Cabbage	Sept.-Oct.	Oct.-Nov.	Dec.-March
Capsicum	Nov.-Jan.	Jan.-Feb.	April-May
	June-July	July-Aug.	Sept.-Oct.
Carrot	Aug-Oct.	-	Dec.-March
Cauliflower	Early June	July	Nov.
	July-Sept.	Aug.-Oct.	Nov.-Jan.
	Sept.-Oct.	Oct.-Nov.	Jan.-March
Chillies	Nov.-Jan.	Jan.-March	April-June
	May-June	June-July	Sept.-Nov.
Clusterbean	Feb.-March	-	April-June
	June-July		Aug.-Oct.
Cowpea	June-July	-	Aug-Oct.
	Feb.-March		April-June
Cucumber	Feb.-March	-	May-July
	June-July		Aug-Oct.
Dolichos bean	June-July	-	Oct.-Dec.
Fenugreek	Sept.-Nov.	-	Nov.-Feb.
Frenchbean	Feb. - March	-	April-May
Kangkong	Feb-Oct	-	Round the year
Lettuce	Sept.-Oct.	Oct.-Nov.	Dec.-Feb.
Mustard	Sept.-Nov.	-	Nov.-Feb.
	Feb-March		March-May
Okra	Feb.-March	-	March-June
	June-July		Aug.-Nov.
Onion	Oct.-Nov.	Dec.-Jan.	April-June
	May-June	June-July	Oct.-Nov.
Garden peas	Sept.-Oct.	-	Nov.-Jan.
	Oct.-Nov	-	Jan.-March
Radish	April-Aug.	-	May-Sept
	Sept-Oct.		Nov.-Jan.
	Nov.-Jan		Dec.-March
Spinach	Sept.-Nov.	-	Nov.-Feb.
	Feb.		March-April
Sponge gourd	Feb.-March	-	April-June
Sweet potato	April-August	-	June-September
Vegetable soybean	June-August	-	July-Sept

Contd...

Crops	Sowing Period	Transplanting Period	Harvesting Time
Ridge gourd	June-July	-	Aug.-Oct.
Round melon	Feb.-March June-July	-	May-June Sept-Oct.
Tomato	June-Aug. Nov.-Dec.	Aug.-Sept. Dec.-Feb	Oct.-Dec. April-June.
Turnip	Oct.-Nov.	-	Dec.-March.
Yard long bean	Feb-Oct	-	April-March

Table 4.3: Monthly Chart for Home Garden Vegetable Cultivation

Month	North India	South India
January	Brinjal	Lettuce,Spinach, Gourds, Melons, Radish, Carrot, Onion, Tomato,Okra,Brinjal, Bean
February	Applegourd, Bittergourd, Bottle gourd, Cucumber, French Beans, Okra, Sponge, Gourd, Watermelon, Spinach	Lettuce,Spinach, Gourds, Melons, Radish, Carrot, Onion, Tomato,Okra,Brinjal, Bean
March	Applegourd, Bittergourd, Bottle gourd, Cucumber, French Beans, Okra, Sponge, Gourd, Watermelon, Spinach	Amaranthus, Coriander, Gourds, Beans, Melons, Spinach, Okra
April	Capsicum	Onion, Amaranthus, Coriander, Gourds, Okra, Tomato, Chilli
May	Onion, Pepper, Brinjal	Okra, Onion, Chilli
June	All gourds, Brinjal, Cucumber, Cauliflower (Early), Okra, Onion,Sem,Tomato,Pepper	All Gourds, Solanaeceae,Almost all vegetables
July	All gourds, Cucumber, Okra, Sem, Tomato	All Gourds, Solanaeceae,Almost all vegetables
August	Carrot, Cauliflower, Radish, Tomato	Carrot, Cauliflower, Beans, Beet
September	Cabbage, Carrot, Cauliflower, Peas, Radish, Tomato, Lettuce	Cauliflower, Cucumber, Onion,Peas,Spinach
October	Beet, Brinjal, Cabbage, Cauliflower, Lettuce, Peas, Radish, Spinach, Turnip	Brinjal, Cabbage,Capsicum,Cucumber, Beans,Peas, Spinach, Turnip, Watermelon
November	Turnip, Tomato, Radish, Pepper, Peas, Beet	Beet, Eggplant, Cabbage, Carrot, Beans, Lettuce, Melon, Okra, Turnip
December	Tomato	Lettuce, Pumpkin, Watermelon, Muskmelon, Ash gourd, Ridge gourd, Bitter gourd, Bottle gourd, Cucumber, Chilly, Cabbage,

4.14.1 Fencing Home Gardens

Live Fencing

Vegetablecrops need protection from animals and sometimes from people. The idea behind a living fence is to use certain plants that make good fence and at the same time produce useful things for people, for livestock and for soil improvement. Goats and chicks that roam freely in villages can cause a great deal of damage to home gardens, there fences should be erected around vegetables. A living fence can be made with sticks of cassava bound with bamboo strips. Another alternative could be using cactus plants or Agave surrounding a home garden. A good living fence can provide fuel wood, animal feed, fruits, building materials, etc.; and may serve as a windbreak and as insects and pests Repellent. Iron net, bamboo sticks, tree sticks and thorny branches, babool, ber, caronda

Table 4.4: Live Fences with Different Benefits and Uses

Fuel wood, timber and basket-weaving	Fruits and other food	Fodder and feed	Mulch and green manure
Acacia spp.	Annona squamosa	*Acacia albida*	*Acacia*
Bamboo	Moringa	*Acacia nilotica*	*Crotalaria*
Eucalyptus spp.	Banana	*Acacia tortilis*	*Gliricidia* spp.
Parkia biglobosa	Cassava	*Artocarpus* spp.	
Prosopis africana	Drumstick	*Gliricidia sepium*	
Pterocarpus spp.	Papaya	*Grewia optiva*	
Raphia palm	Passion fruit	*Morus* spp.	
	Pigeon pea	*Morus* spp.	
	Yam	*Ziziphus*	
	Cassava	*Gliricidia* spp.	

4.14.2 Spacing, Mulching, Staking, Pruning and Thinning

i. Spacing: Ideal spacing and plant population are for maximizing yield and quality without unduly increasing cost. All crops tend to increase yield per unit area as plant population increases, but only up to a certain limit as the yield may not increase beyond a limit and may even drop. Appropriate spacing differs from situation to situation.

- Population range for okra is 30,000 to 120,000/ha, depending primarily on the variety.

- Cabbage is planted at a higher density if smaller heads are preferred.
- There is a lot of flexibility for spacing within rows and, consequently, for plant population.

Table 4.5: Planting Guide: Spacing

Vegetable	Inches Between Plants	Inches Between Rows	Depth of Seed (Inches)
Asparagus	12-18	36-48	1-1 ½
Beans, broad	8-10	36-48	1-2
Beans, dry	4-6	18-24	1-1 ½
Beans, lima			1-1 ½
Bush	2-3	18-24	
Pole	4-6	30-36	
Beans, mung	18-20	18-24	½
Beans, snap/green Bush			1-1 ½
Pole	2-3	18-24	
	4-6	30-36	
Beets	2-3	12-18	1-1 ½
Broccoli	3	24-36	1
Brussels sprouts	24	24-36	½
Cabbage	18-24	24-36	½
Carrot	2-4	12-24	¼
Cauliflower	18-24	24-36	½
Celery	8-10	24-30	¼
Chayote	24-30	60	
Chinese cabbage	8-12	18-30	½
Corn	2-4	12-18	1-1 ½
Cress	1-2	18-24	¼
Cucumber	12	18-72	½
Eggplant	18-24	24-36	¼
Kale	8-12	18-24	½
Kohlrabi	5-6	18-24	¼
Leek	6-9	12-18	1/8

Contd...

Vegetable	Inches Between Plants	Inches Between Rows	Depth of Seed (Inches)
Lettuce	6-12	12-18	1/8
Muskmelon	18-24	60-96	1
Mustard	6-12	12-24	½
Okra	12-18	24-36	½-1
Onion Sets	2-3	12-18	1-2
Onion Seeds	1-2	12-18	1/4
Pea, shelling	1-2	18-24	2
Pepper	18-24	24-36	½
Potato, Irish	12-18	24-36	4
Pumpkin	24-48	60-120	1
Radish	1-6	12-18	½
Shallot	6-8	12-18	¼
Sorrel	12-18	18-24	½
Soybean	11/2-2	24-30	½-1
Spinach	2-4	12-24	½
Spinach, New Zealand	12	24-36	½
Squash, summer	24-36	18-48	1
Squash, winter	24-48	60-120	1
Sweet potato	12-18	36-48	3-5
Tomato	18-36	24-48	½
Turnip Greens	2-3	12-24	½
Turnip Roots	3-4	12-24	½
Watermelon	24-72	60-120	1
Coriander	12	12-18	¼
Basil	4-6	18-24	¼
Garlic	3-6	24-36	¼
Mint	2-3	18-24	¼
Oregano	6-12	12-18	¼
Parsley	12-18	18-24	¼

Contd...

4.15 Staking and Training

All climber vegetables are staked. There are three types of plants for the purpose of staking: Plants (like cucurbits) with special structures such as tendrils which allow them to climb;Plants that twine (such as yard-long beans); and Plants (such as tomatoes) that do not have natural ability to climb and must, therefore, be tied to stakes.

Fig. 4.8: Staking Ridge-Gourd and Tomato

Advantage: Staking facilitates management operations such as irrigation, inter-tillage, pest control and harvesting. It also helps produce better product.

Disadvantages: It can be very costly, accounting for as much as 50% of production cost. However, where labour and materials for staking are cheap, it would be preferrable to stake. Training or repositioning of vines is done to crops, which are grown under prostrate culture (without stakes) to prevent overcrowding in some spots of the field. In insect-pollinated crops, such as watermelon and squash, dense vines and foliage may interfere with insect activity and reduce fruit- set. In staked crops, training is necessary in the initial stages to keep vines off the ground.

4.15.1 Pruning

In indeterminate tomatoes, pruning results in single-stem plants which can easily be tied to the stake. Thus the fruits are larger because plant's nutrients are not diverted to branches. In the case of cucurbits, such as *Luffa*, pruning of the tip of the seedling stimulates early branching and fruiting on lower nodes. A distinct method of pruning is done on mature plants ,which show declined productivity. The main stem is cut 20 cm from the ground and stimulated to produce new branches by applying fertilizers and irrigation. The result is a ratoon crop, which starts producing fruits sooner than seed planting. However, the yields are usually lower. The practice of rationing works very well with some varieties of okra and eggplant during the wet season. Pruning should be done with a sharp blade to minimize damage to plant tissues and to facilitate recovery. To prevent spread of diseases, the blade should be dipped regularly in detergent solution while being used.

4.15.2 Thinning of Seedlings

Thinning provides better environment for the remaining seedlings by giving proper space for good growth and development. Some of the reasons for thinning are as follows.

- It reduces competition among seedlings for soil nutrients and water.
- It can reduce some early disease problems by providing better air circulation around the plants.
- It provides conditions that are more nearly ideal for growing healthier vegetables of desired size, shape and weight.
- Vegetable plants that are properly spaced produce higher yields.

Start thinning when plants have one or two pairs of true leaves. The ideal time for thinning is when the soil is damp and soft but not excessively wet. If the soil has become dry, water the site for an hour or more before thinning is to be started. Evening is a good time to thin because the remaining plants have cool, dark night to recover. Practically every vegetable that is directly seeded will benefit from thinning. However, thinning is a must for lettuce, beets, radishes, carrots, spinach and turnips.

4.15.3 Fruit Thinning

To control fruit size, some fruits are removed before they start growing. Some plants particularly cucurbits, produce female flowers and set fruit so early that vegetative growth is still insufficient to support normal growth of the fruits. When this happens, further vegetative growth is restricted, while additional fruit setting and development is equally affected. To promote the formation of bigger and better fruits, the first one or two fruits on the vine need to be removed. The number of fruits per vine subsequently are limited to one. The practice of fruit thinning is widely used in melons and watermelons.

4.16 Protected Cultivation

The need for a year-round supply of fresh vegetables cannot be met adequately in many parts of the world, considering that there are cycles in seasons and unpredictable weather. Even during the most favourable season, there are environmental constraints in production , limiting yield and quality of vegetables. These are compounded by detrimental effect of pesticide residues on vegetables. This awareness has led to the development of production systems that provide physical barriers to unfavourable climatic and biological influences, called protected cultivation.

Protected cultivation was started in the temperate zones, primarily to produce vegetables during cold winter months or prolong growing season by starting early in spring and extending up to late fall. This type of production system has become a distinct system in cold countries.

In the tropics, the most favourable season for growing many vegetables is winter season. Besides winter season, temperature becomes too high for some vegetables and precipitation too heavy for many. Some growers have found it profitable to adopt protected cultivation techniques even though at a greater expense. The method may vary from the simplest net enclosure to sophisticated plastic houses with movable roofs. The use of net enclosures is primarily aimed at controlling insect -pests that are difficult to manage by intensive use of pesticides. The most common crops that are protected by nets are tomato, cucumber, pepper, the Brassicas, especially cauliflower and broccoli, the high -value crops, which are highly susceptible to diamondback moth. The nets also tend to reduce temperature in the field and break raindrops into sprays. The use of plastics in the tropics is primarily aimed at protecting crop against excessive rain. The main problem of plastics in the tropics is the build-up of heat inside the enclosure. However, with improvements in plastic house designs, it has become possible to use plastics even in warm climate of low elevations.

4.17 Weed Control in Home Gardens

A weed may be broadly defined as any unwanted plant, or a plant out of place. Most weeds are plants that are generally considered undesirable by gardeners. They are neither eaten nor considered attractive. They are frequently invasive and difficult to control. Weeds compete with crop- plants for moisture, nutrients and light. They may also harbour insects ,which would harm vegetables or flowers or transmit diseases to crop. Weeds can also serve as alternate hosts for diseases. They also promote diseases by increasing humidity, decrease vegetable quality and make difficult even harvesting vegetables. The principle methods of weed control fall into three categories: mechanical, cultural and chemical.

Fig. 4.9: Hand Weeding in Home Garden

4.17.1 Mechanical Methods

It involves removing weeds while they are small and stopping them to grow and produce mature seed. It consists principally of mowing, plowing, roto-tilling, hoeing and hand- pulling.

Disadvantages: It works only on growing weeds and may need to be repeated frequently. Remove weeds while they are small because it is quicker, easier and does less damage to desired vegetable -plants. Hoe or till shallowly (less than 5- cm deep) to avoid damage to the desirable plants and to minimize moisture loss from the soil. Most weed seeds ,which germinate are in the upper 5 cm of soil. Gardeners often quit weeding as crop matures. Weeds do not affect crop yields as much at this stage. Weed removal usually should continue until the vegetable crop is harvested. When crops are not growing on the garden spot, the soil may then be kept tilled, mowed or heavily mulched to stop growing of weeds.

Frequent tillage has additional advantages of turning organic material and mechanically destroying any insects that are present.

4.17.2 Cultural Methods

Most cultural methods of weed control emphasize prevention. These include mulching, solarization and multiple or wide-row planting.

a) Mulches

Any layer of material spread over the soil surface may be considered a mulch. There are two classes of mulches: organic and inorganic. Both reduce weed growth, retain soil moisture and influence soil temperature. Organic mulches include ground corn-cobs, shredded leaves, grass clippings, straw, pine- needles, wood chips, shredded bark, hay, sawdust, ground bark, compost and even newspapers. All reduce weed emergence and make it easier to pull those weeds that do emerge. Inorganic mulches include black and other opaque plastic as well as landscape fabric. Landscape fabric is much more expensive than plastic mulch and allows water to pass through into the soil. Black plastic (polyethylene) is the most commonly used inorganic mulch. Clear or translucent plastic should not be used as mulch. These plastics allow light to penetrate, which permits germinated weed seeds to grow even under plastic mulch. Black plastic eliminates growth of most, but not all weeds. Opaque plastic in colours other than black is sometimes used as mulch, which also controls weeds. However, many of these plastics are quite expensive. Due to high temperature and intense solar radiation, use of black plastic may cause damage to plants transplanted or seeded during the hottest part of the summer. White plastic can be utilized during the summer months to prevent root and stem damage to the plant. White plastic would allow penetration of light. Therefore, white plastic with a black backing (known as white on black) is preferable.White on black plastic is more expensive and may be difficult for home gardeners.

Fig. 4.10: Straw Mulch Tomato and Plastic Mulch in Brinjal

b) Solarization

It is the use of clear polyethylene plastic sheeting (2 to 6 mils thick) to capture radiant energy of the sun, thereby raising soil temperature to levels lethal to many weed seeds. The plastic sheeting is placed over bare, moist soil during a summer fallow period. The top layer of the plastic should be kept clean to maximize solarization efficiency(Fig11). An occasional dusting with a dust- mop or rinsing with a garden -hose should be all that is a necessity. The plastic is left in place for 4 – 5 weeks. During this period, soil temperatures are frequently raised to 48 – 52ºC. This practically eliminates viable weed seeds in the top 5-7.5 cm of soil. After the plastic is removed, care must be taken not to mix deeper layers of un-solarized soil with the nearly weed-free solarized soil. Many soil-borne plant pathogens would also be reduced during solarization. This improves plant standand vigour and may double yields.

Fig. 4.11: Soil Solarization Treatment for Weed Management in Net-house for Vegetable Cultivation

c) Multiple or wide row planting

Closely spaced double or triple rows may also assist in weed control. Simply plant two or three rows of a vegetable close enough so that leaves cover the area between them rapidly as plants grow. Very small vegetables such as radishes may be broadcast in a long row 30 cm or so wide. These techniques allow growing vegetables to shade soil, which reduce weed growth.

d) Chemical methods

Chemicals (herbicides) are only occasionally used by home gardeners to control weeds because suitable herbicides are seldom legally available in small, economical amountsMany herbicides are nonselective and would also kill vegetables, besides weeds.Others are selective and can be used only with certain vegetables or control only certain weeds. Home gardens generally contain many vegetable and weed species, which makes using selective herbicides difficult. Some herbicides may also damage nearby vegetables or remain in the soil and damage future plantings. The herbicide application rate may be very low and extreme accuracy in application may be absolutely essential. Overlapping applications may kill vegetable crops and, if areas are skipped, weeds may not be controlled.Required pre-harvest intervals (PHI's) or waiting periods between application and harvest can be lengthy and must be observed. Herbicides may also be effective only for a short period of time or produce results slowly. Two herbicides that may be helpful in the home garden are Preen and Dacthal (DCPA). Both products are pre-emergence herbicides. They must be applied before the germination of weed seeds. Dacthal is most effective on annual grasses. When selecting pre-emergence herbicides, see the herbicide label for the list of vegetables and ornamentals for which it has been approved. If a plant species does not appear on the herbicide label, it is not legal to use it on that plant.

A post-emergence herbicide that can be used in the home garden is glyphosate (Roundup). Glyphosate is a nonselective, systemic herbicide. It is absorbed by green, actively growing plants and translocate throughout the entire plant. Glyphosate kills virtually all plants that are directly sprayed. Read the label directions carefully for specific recommendations.

5

Vegetables in Nutrition

The home garden is the common type of nutrition garden. Its prime function is to provide a supplementary source of essential nutrients for the family diet. Three nutritional problems that have serious consequences include deficiencies of iron, vitamin A and iodine. South Asia has the highest rate of malnutrition, making it the single largest cause of child mortality and blindness in the region. Nutritional security is more than just food security. It is the outcome of good health, a healthy environment and good caring practices. However, despite good economic growth in several South Asian countries, there has been insufficient progress in reducing under-nutrition in recent years. Iron deficiency anemia (IDA) continues to be major public health problem. Inadequate dietary intake of iron, defective iron absorption, increased iron requirements due to repeated pregnancies and lactation, frequency of infections in children, and excessive physiological blood loss during adolescence and pregnancy are some of the causes responsible for the high prevalence of anemia in developing countries.WHO estimates that one out of every two preschool children and pregnant women in developing countries are iron deficient. The problem is evidently linked to poverty lucent girls are severely anemic. .

Horticultural crops provide a wealth of nutrient. It needs only determination to popularize and use them for better nutrition. Vegetables are a source of energy, vitamins, minerals and green leafy vegetables are good sources of Vitamin B complex and C. Their description and importance is given below.

5.1 Types of Nutrients

5.1.1 Calories

In general, higher the water content, lower is the calorie content. Roots, tubers and seeds of plants have a higher sugar and starch content and less water than other parts. Thus, these provide more calories per unit weight.

5.1.2 Protein

Vegetables are generally a poor source of protein except legumes. Two vegetables which have good quality protein include: potato and soybean, grain legumes or pulses contain 20-40% protein.

5.1.3 Minerals

Vegetables are a rich source of both calcium and iron; the two most important minerals needed by the body. Calcium is needed for the development and proper functioning of bones and teeth while Iron is needed to prevent anemia.

5.1.4 Vitamins

Vegetables are excellent source of vitamins A, B complex and C. Lack of vitamin A causes poor growth and night blindness which develops into xerophthalmia (deterioration of eyesight) and ultimately leads to complete blindness. Dark green and yellow vegetables are rich in provitamin A, the precursor of Vitamin A. Young shoots contain more of this vitamin than the mature leaves. Mortality caused by lack of vitamin A accompanied by protein and calorie malnutrition may be as high as 80%.

Lack of vitamin C causes scurvy a disease of the gums causing sponginess and bleeding. Single helping of vegetables (~ 91 g daily) provides 10mg of ascorbic acid, an amount known to prevent scurvy, increase resistance to colds, cough and other respiratory diseases. Green leafy vegetables are good sources of Vitamin B and C.

5.1.5 Fiber

Fiber not a nutrient is absorbed by the body, aids in muscular action of the intestines, satisfies appetite and prevent constipation. Use of diet rich in fiber controls blood cholesterol level and high blood pressure, prevents heart , gall stones and colon cancer and prevents and treats obesity and controls diabetes.

5.2 Role of Vegetables in Health and Nutrition

Vegetables are powerful allies in the fight against disease. Studies show that the nutrients and antioxidants in vegetables can help prevent and treat cancer, asthma, heart disease and many other chronic illnesses.

5.2.1 Cancer

Experts believe that one third of all cancers could be prevented by a proper diet. The most common types of cancers are carcinomas found in breast, prostate, lung, colon, and skin. Many types of vegetables help prevent cancer, among other diseases. Numerous studies show that spinach can lower the incidence of colon, lung, skin, oral, stomach, ovarian, prostate, and breast cancers. While spinach is used in most studies, other leafy greens with similar nutrient profiles, such as kale, collards, Swiss chard, mustard greens, turnip greens, lettuce, and orange bell peppers have similar cancer-fighting properties. Other vegetables shown to be excellent cancer-fighters include soybean, cabbage, carrot, and parsnip. Vitamin A, found in orange and leafy green vegetables, inhibits the growth of carcinomas. Carotenoids in red and yellow vegetables, as well as vitamin C in brassicas, tomato and pepper, play a role as well.

Phytoestrogens, found in soy foods, can help prevent prostate and breast cancers. Evidence is also mounting that eating phytoestrogen-rich beans and legumes of all types may help prevent pancreatic, colon, breast, and prostate cancer. The most-touted cancer-fighting vegetable is broccoli, which is rich in sulfur compounds including sulphuraphane - which actually kills cancer cells - and indoles, which inhibit the growth of breast cancer tumours. Strong-flavoured vegetables with sulfurous cooking odours - including cabbage, Brussel's sprouts, mustard greens and horseradish - are rich in these sulfur-containing compounds, which are called glucosinolates. Broccoli is also a great source of vitamin C, beta-carotenes, and fibre: all cancer-fighters. While broccoli is the most studied, other members of the brassica vegetable family are also beneficial. One study has found that onions and other alliums - especially shallots - which are rich in antioxidants called phenolics and flavanoids were extremely effective against liver and colon cancer cells. Pungent allium varieties such as yellow onions contain more of the cancer-fighting phytochemicals.

5.2.2 Heart Disease and Diabetes

Flavonoids in onions and alliums have also been shown to have anti-bacterial, anti-viral, and anti-inflammatory properties which reduce the risk of heart disease and diabetes. All forms of beans and legumes are great for a healthy heart - just half a cup of legumes per day will keep heart healthy and contribute to it being disease-free. Several studies show that people who eat more beans have lower cholesterol, lower blood pressure, and a significantly lower risk of heart disease and a much lower incidence of diabetes. Other vegetables that help fight heart disease include broccoli, garlic, kale, and soy foods.

5.2.3 Asthma

Studies show that vitamins A, C, E, and B6, as well as the phytochemicals in vegetables, help alleviate the airway inflammation and damaged lung tissue found in asthma and respiratory disease. Calcium is also beneficial. Magnesium has even been shown to be effective in treating severe, acute asthma attacks, contributing to respiratory health. Research also shows that children who eat foods rich in omega-3 fatty acids are less likely to have asthma than those with diets high in omega-6 fatty acids. Omega-3 fatty acids reduce inflammation, while omega-6 fatty acids increase inflammation. Vegetable oils that contain omega-3 fatty acids, such as canola or olive oil, should be preferred over the vegetable oils or margarine containing omega-6 fatty acids.

5.2.4 Arthritis

Vegetables rich in antioxidants can help fight arthritis. Researchers have found that people with diets high in the carotenoids beta-cryptoxanthin and zeaxanthin were less likely to develop inflammatory arthritis. People who developed arthritis had a 40 percent lower intake of the antioxidant beta-cryptoxanthin than those who didn't develop the disease. Yellow and orange fruits and vegetables, including bell peppers and corn, are good sources of these antioxidants.

5.2.5 Osteoporosis

Studies show a diet rich in fruits and vegetables, as well as calcium, magnesium, and potassium, increases bone density. Phytoestrogens in soy products have also been shown to strengthen bones in post-menopausal women. Soy products, such as tofu or soy milk, are also good for menopausal symptoms.

5.2.6 Macular Degeneration and Cataracts

Eating foods rich in the carotenoids lutein and zeaxanthin - including spinach, kale, collards, turnip greens, orange bell peppers, and corn - decreases the incidence of macular degeneration and cataracts. Omega-3 fatty acids - including canola and olive oil - are also beneficial in preventing eye disease.

5.2.7 Colds and Environmental Toxins

Zinc is a powerful antioxidant that has been shown to be effective in fighting infections such as colds, and environmental toxins, such as air pollution.

5.2.8 Emotional and Mental Health

Eating dark leafy greens and legumes could actually make you happier. People who suffer from depression have been shown to be deficient in B-complex vitamins.

Folic acid (B9) is used to treat depression, and niacin (B3) can help anxiety. Behavioral problems such as poor self-control, anger, rage, dementia, and Alzheimer's have been linked to vitamin B12 deficiency. The B-complex vitamins, iron, and trace minerals - especially zinc - are important for good mental health. The best way to fight such diseases is with proper nutrition. Eat a wide variety of whole, unprocessed foods - including lots of vegetables for a healthy mind and body!

5.3 Colored Vegetables

Eating a rainbow of fruits and vegetables is a simple way of remembering to get as much of variety color in your diet as possible, so that you can maximize your intake of a broad range of nutrients. The colors of fruits and vegetables offers a small clue as to what vitamins and nutrients these contain. By getting a variety of different colored fruits and vegetables, you ar guaranteed a diverse amount of essential vitamins and minerals.

5.3.1 Red

Contain nutrients such as lycopene, ellagic acid, quercetin, and hesperidin, to name a few. These nutrients reduce the risk of prostate cancer, lower blood pressure, reduce tumor growth and LDL cholesterol levels, scavenge harmful free-radicals, and support joint tissue in cases of arthritis. This includes; Beetroot, radish, red bell peppers, red chili peppers, red onion, tomato, African eggplant, beet leaf.

5.3.2 Orange and Yellow

Contain beta-carotene, zeaxanthin, flavonoids, lycopene, potassium, and vitamin C. These nutrients reduce age-related macula degeneration and the risk of prostate cancer, lower LDL cholesterol and blood pressure, promote collagen formation and healthy joints, fight harmful free radicals, encourage alkaline balance, and work with magnesium and calcium to build healthy bones. These include: carrot, sweet corn, sweet potato, yellow bell pepper, squash, pumpkin, yellow summer squash, Yellow tomatoes, yellow winter squash, zucchini.

5.3.3 Green

Contain chlorophyll, fiber, lutein, zeaxanthin, calcium, folate, vitamin C, calcium, and Beta-carotene. The nutrients found in these vegetables reduce cancer risks, lower blood pressure and LDL cholesterol levels, normalize digestion time, support retinal health and vision, fight harmful free-radicals, and boost immune system activity. These include: asparagus, broccoli, brussel's sprouts, celery, chayote, squash, chinese cabbage, cucumber, beans, green cabbage, green onion, green peppers, leafy greens, leeks, lettuce, okra, garden pea, snow pea, spinach, zucchini.

5.3.4 Purple

Contain nutrients which include lutein, zeaxanthin, resveratrol, vitamin C, fiber, flavonoids, ellagic acid, and quercetin. These nutrients support retinal health, lower LDL cholesterol, boost immune system activity, support healthy digestion, improve calcium and other mineral absorption, fight inflammation, reduce tumor growth, act as an anti-carcinogens in the digestive tract, and limit the activity of cancer cells. These include: brinjal, purple coloured asparagus, cabbage, carrots, peppers, dolichos bean, and sword bean.

5.3.5 White

Contain nutrients such as beta-glucans and lignans that provide powerful immune boosting activity. These nutrients also activate natural killer B and T cells, reduce the risk of colon, breast, and prostate cancers, and balance hormone levels, reducing the risk of hormone-related cancers. These include: Cauliflower, onion, garlic, ginger, kohlrabi, shallot, turnip, radish, agathi.

5.4 Nutritional Elements and their Role with Respect to Vegetables

a) Vitamin A (Retinol)

is important for the health of bones, teeth, skin, and eyes; keeps the mucous membranes of the lungs and nose moist and is important for respiratory health. Vegetables which are good sources of vitamin A - many of which are orange or dark green – e.g., broccoli, carrots, dark leafy greens, tomato, pumpkin, sweet potato, winter and summer squashes, beetroot and garlic.

b) B-Complex Vitamins

The B vitamins (also called B-complex) help convert carbohydrates into sugar, thus providing energy to the body. They also help process fat and protein, tone muscles in the gastrointestinal tract, and aid the nervous system.

c) Vitamin B$_1$ (Thiamin)

is good for the heart, muscles, and nervous system. If one feels tired and run down, he may not be getting enough thiamin. Green pea, corn, legumes, and leafy greens are good sources of thiamin.

d) Vitamin B$_2$ (Riboflavin)

helps the body to use oxygen and is an antioxidant. Legumes, broccoli, spinach, asparagus, and leafy greens such as spinach and turnip greens are good sources of vitamin B$_2$.

e) Vitamin B$_3$ (Niacin)

is needed for a healthy nervous system and gastrointestinal tract, as well as healthy skin. Protein-rich foods contain vitamin B3. Good vegetable sources are leafy greens, asparagus, legumes, potato, and peanut.

f) Vitamin B$_5$ (Pantothenic Acid)

is necessary to produce steroids and cortisone in the adrenal gland. Often called the anti-stress vitamin, it strengthens the immune system and helps the body deal with stress. Pantothenic acid is found in a wide variety of foods - in fact it gets its name from the Greek pantos meaning everywhere. Good vegetable sources include corn, legumes, brassicas, tomato, and sweet potato.

g) Vitamin B$_6$ (Pyridoxine)

is needed to produce red blood cells. Vitamin B6 is found in many foods including tomatoes, potatoes, and soybeans.

h) Vitamin B$_9$ (Folate or Folic Acid)

is required for creating and maintaining healthy cells, cell division, and the formation of hemoglobin in red blood cells. Sometimes called the brain vitamin, it's necessary for mental and emotional health. Good vegetable sources of folates are dark leafy greens, root vegetables, soybeans and other legumes.

i) Vitamin B$_{12}$

is important for cell function, the formation of red blood cells, healthy bone marrow, and healthy nerve sheaths. Lack of vitamin B12 can cause anaemia. Vitamin B12 cannot be manufactured either by plants or animals, but can only be synthesized by bacteria, which are naturally present in the human small intestine as well an in meats, shellfish, and dairy products. There are no vegetable sources of vitamin B12. Vegetarians may get vitamin B12 from foods containing yeast, such as Marmite, or fermented foods such as tempeh. Lacto-ovo vegetarians may also obtain the B-complex vitamins through eating dairy products. Vegans are often advised to take a B-complex supplements.

j) Vitamin C

is an antioxidant important in preventing infection, healing wounds, and promoting healthy bones and teeth. Vegetables that are good sources of vitamin C include cruciferous vegetables such as broccoli, cauliflower, and cabbage; leafy greens such as collards, spinach, bell peppers and tomatoes.

k) Vitamin D

is important for strong bones and teeth, blood clotting, and the absorption of calcium and magnesium. There are no vegetables that contain significant amounts of vitamin D. The best source of vitamin D is sunlight. The body manufactures its own vitamin D from exposure to sunlight: 10 to 15 minutes of sunlight a day provides all the vitamin D the body needs.

l) Vitamin E

is an antioxidant that is beneficial to the immune system, cell membranes, and body tissues. The best sources of vitamin E are vegetable oils such as canola, corn, peanut, and soybean, as well as dark leafy greens, legumes, broccoli, and pumpkin.

m) Vitamin H (Biotin)

is needed for healthy hair, skin, and nails. Good vegetable sources of biotin are legumes, such as soybean and peanut besides green vegetables. Studies have shown that vegetarians actually absorb more biotin from the gastrointestinal tract than people who eat meat.

n) Vitamin K

helps blood to clot (the "K" comes from the German word koagulation) and is important in regulating calcium levels and promoting bone health. Vegetables that are good sources of vitamin K include leafy greens, such as spinach and lettuce, and cruciferous vegetables such as broccoli, cauliflower, and cabbage.

5.5 Minerals

a) Calcium

is the body's most abundant mineral. It's needed for strong teeth and bones, as well as muscle and nerve health. Good vegetable sources of calcium are broccoli, cabbage, dark leafy greens, and legumes.

b) Phosphorus

is the body's next most abundant mineral. It's important for bones and teeth, as well as energy. Phosphorus is contained in many foods. Good vegetable sources of phosphorus include legumes, potato, and garlic.

c) Magnesium

is important for all of the body's organs, especially the heart, kidneys, muscles, bones, and teeth. It also activates enzymes. It is found in whole foods. Many people do not get enough magnesium, since it's hard to get enough of the mineral with a diet of processed food. Good vegetable sources of magnesium are dark leafy greens, legumes, tofu, peanuts, and unpeeled potatoes.

d) Potassium

is needed for kidney function, as well as for the heart, digestive system, and muscles. Potassium is present in many vegetables, especially potato, broccoli, spinach, and chick peas.

e) Iron

is essential for blood cells and helps in the delivery of oxygen to the body's tissues. Lack of adequate iron can cause low energy and anemia. Good vegetable sources include dark leafy greens and legumes.

f) Zinc

is important for the immune and reproductive systems. It is also an antioxidant, and can help fight colds and environmental toxins. Zinc also acts to regulate the appetite, and contributes to the senses of taste and smell. Vegetable sources of zinc include legumes (particularly limas, pintos, soybeans, and peanuts), tofu, string beans, leafy greens, and pumpkin seeds.

5.6 Sources of Vitamins

While all vegetables contain vitamins, minerals, antioxidants, and other nutrients, the vegetables profiled here are veritable treasure of nutrition.

a) Beans, Peas and Legumes

Inexpensive and versatile, beans are packed with B vitamins, especially folic acid, niacin, riboflavin, and thiamin, as well as magnesium, potassium, and phytonutrients. They're also a good source of protein, fibre, and antioxidants. Dried beans contain more of water-soluble vitamins than canned beans, but both are good. Beans are part of the Legume family, which includes string beans, lima beans, peas, lentils, chick peas (garbanzos), soybeans, peanuts, and all forms of dried beans.

Daily consumption of green peas along with other legumes lowered the risk of stomach cancer.

Vegetable soybeans are a complete protein, a good meat substitute, and a great source of vitamin E, potassium, folic acid, magnesium, and selenium. In addition, it is rich in phytoestrogens, compounds which mimic the human oestrogen hormone. Common forms of soy include tofu, soy milk, soy cheese, soy cold cuts, soy sausage, tempeh, edamame, and soy flour. A half a cup of tofu contains about 40 percent of protein, 25 percent of calcium, and almost 90 percent iron needed by an adult woman. Soy is also a good dairy substitute for people who are lactose intolerant.

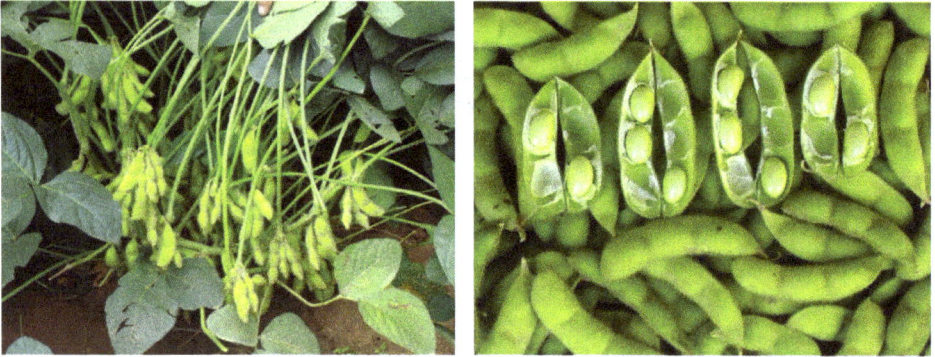

Fig. 5.1: Vegetable Soybean Plant and its Eatable Pods

b) Broccoli and the Brassicas

Broccoli is high in vitamin C, vitamin K, beta-carotenes, calcium, and iron. One cup of broccoli provides 100 percent daily requirement of vitamin C and it's a great source of iron. Broccoli also supplies significant amounts of vitamin E, vitamin B6, and fibre. Broccoli is a member of the Brassica family, which also includes, cabbage, and many dark, leafy greens such as collards, kale, Swiss chard, and mustard greens. Broccoli is full of cancer-fighting antioxidants.

c) Spinach and Leafy Greens

Spinach is the most popular of the dark, leafy greens, but collards, kale, mustard greens, and Swiss chard are also packed with the same nutrients. Spinach is rich in vitamins C and E, beta-carotene, lutein, the B vitamins (thiamin, riboflavin, B6, and folic acid), omega-3 fatty acids, calcium, iron, magnesium, manganese, zinc, and many other micro- and phytonutrients. Spinach is packed with antioxidants that promote healthy eyes and help prevent macular degeneration, the leading cause of blindness in older adults. Cooking the green helps make lutein (α-a carotenoid) more absorbable by your body.

d) Tomato

Tomatoes are rich in vitamin C and lycopene. Lycopene is a powerful antioxidant and is one of the carotenes that give tomatoes their bright red colour. (yellow and orange tomatoes do not contain lycopene, but are still good sources of vitamins and minerals). Tomatoes are one of the few sources of dietary lycopene: other sources include watermelon and grapefruit. Try to eat tomatoes or another lycopene-rich food daily.

e) Potato

Another member of the nightshade family, potatoes are a good source of complex carbohydrates - needed for energy - as well as fibre, vitamin C, and potassium. They also contain many minerals and B-complex vitamins (especially vitamin B6). One potato (with skin) supplies 45 percent of the daily requirement of vitamin C, and more potassium than a banana or a serving of broccoli.

f) Winter Squash and Pumpkin

The pumpkin is a powerhouse vegetable that's often overlooked. Rich in alpha-carotenes, beta-carotenes, vitamin C, vitamin E, pantothenic acid, potassium, magnesium, and fibre, it's a great source of nutrition. Fresh, tinned, or frozen, the orange winter squashes - including pumpkin, butternut, buttercup, and hubbard - should be a mainstay of every diet.

g) Sweet Potato

It is another orange vegetable rich in vitamins A and C, as well as potassium and fibre.

h) Peppers

Bell peppers are rich in vitamins A and C, as well as potassium. A ripe pepper (red, orange, yellow, etc.) contains twice as much vitamin C and ten times as much vitamin A as an unripe green pepper (although all peppers are good) One cup of raw red peppers contains almost 300 percent the necessary daily amount of vitamin C and 100 percent of vitamin A and warding off atherosclerosis, which can lead to heart disease. Orange peppers are an especially good source of the carotene lutein – important for eye and skin health.

i) Onion

Regular doses of onion skin extract helped lower blood pressure for adults with hypertension. The key player is likely quercetin, a type of antioxidant that's also found in tea and berries. Onions are also pretty smart for those watching their weight because an entire large one is just 63 calories and brimming with vitamin C. Cancer-fighting antioxidants by eating it raw; cooking onions at a high heat significantly reduces the benefits of phytochemicals that protect against lung and prostate cancer. Try combining chopped raw onions with tomatoes, avocado, and peppers for a blood sugar friendly chip dip. Finish with a splash of lime juice.

Sweet corn: On the cob or off, just make sure to eat corn as cooked! The longer the corn was cooked, the higher the level of antioxidants like lutein, which combats blindness in older adults

j) Kale

This veggie's curly green are chock full of vitamin C, an antioxidant that may reduce the risk of heart disease by lowering levels of LDL, or "bad" cholesterol.

k) Basil

Fragrant basil is a must-have in gardens for those who love to cook. Basil requires a sunny spot-and really takes off in hot weather. Grow this tasty herb in garden beds or in containers.

l) Brussels Sprouts

These balls of antioxidants can help detoxify cancer-causing free radicals, and with 80 percent of your daily vitamin C in just 1/2 cup, also help fight heart disease and ward off cataracts.

m) Beet

Roasted or pickled, this root vegetable contains high levels of antioxidants that fight cancer, as well as lutein, which protects the eyes. Beet greens are the most nutritious part of the vegetable and can be cooked like any other dark leafy green.

n) Carrot

One cup of sliced carrots contains 50 calories, 3 grams of fiber, a decent amount of vitamin C, and over 400% of your daily vitamin A needs.

o) Pok Choy

1 cup of Pok choy contains only 10 calories, but gives a good dose of fiber and vitamins A and C. Eating cruciferous veggies may also help prevent prostate cancer as well as colorectal cancer.

p) Parsley

Parsley is loaded with vitamins and minerals, including a decent amount of folate. This vitamin B has long been touted as an important nutrient for pregnant women. Sufficient folate intake could reduce the risk of heart disease. Two full cups for just 35 calories.

q) Celery

One cup of chopped stalks is only 16 calories and provides a decent amount of vitamin A and folate. Celery is an excellent source of vitamin K, a key nutrient for bone and blood health.

r) Cabbage

Eating 1 cup of chopped, raw cabbage will only cost you 17 calories and you'll get a good dose of fiber and vitamin C. Cabbage is also a member of the cruciferous vegetable family which contain substances called glucosinolates that have been linked to a reduced risk of cancer.

s) Radish

A full cup contains 19 calories plus a good amount of vitamin C and fiber. Radish are also filled with phytosterols, a type of compound that can help lower cholesterol.

5.7 Nutritional Classification of Vegetables

Vegetables are classified in different ways. Nutritionally they are classified as follows according to the major nutrients they provide:

5.7.1 Vegetable Legumes

Pulses or legumes are rich sources of protein in our diets. In a vegetarian diet or a diet containing low amounts of animal food, they are an important source of protein. Some of the major pulses are green gram and cow pea. In amounts used, pulses and legumes do not contribute much to the total mineral intake. However being rich in B-vitamins, they can contribute significantly to B-vitamin intake. Like cereals, they do not contain any vitamin A or vitamin C but germinated legumes contain some vitamin C. The protein of pulses / legumes is of low quality since they are deficient in tryptophan also. However they are rich in lysine. Hence, they can supplement proteins of cereals and the quality of the protein from a mixture of cereals and pulses is superior to that of the either one. The most effective combination to achieve maximum supplementary effect is 4 parts of cereal protein + 1 part of pulse protein. In terms of the grains, it will be 8 parts of cereals and 1 part of pulses.

5.7.2 Green Leafy Vegetables (GLV)

The commonly consumed greens are palak, amaranth, fenugreek, drumstick, mint etc. The green leafy vegetables are rich source of calcium, iron and beta carotene, vitamin C, riboflavin and folic acid. These greens are inexpensive and it is advisable to include at least 50g of GLV daily in one's diet. They contain all important nutrients required for growth and maintenance of health. Hence GLV must be consumed by children, pregnant and nursing women to obtain much needed beta-carotene, calcium and iron. This is particularly so on a predominantly cereal based diets of the poor who suffer from the dietary deficiency of these nutrients. Hence steps must be taken to encourage cultivation of GLV in home gardens and school gardens so that they are available all through the year. Use of green leaves from trees like drumstick, agathi etc. can be obtained, regularly without much effort if a tree is planted in the backyard.

5.7.3 Roots and Tubers

Some of the important root vegetables which are commonly consumed are cassava, potato, sweet potato, carrots and colocasia. These are all rich in carbohydrate and can form an important source of energy in our diets. Carrot is rich in carotene and potato is a significant source of vitamin C. Cassava is rich in calcium. Cassava is consumed in Kerala supplies energy and helps to meet a short supply in cereals during drought conditions.

5.7.4 Other Vegetables

Vegetables other than green leafy vegetables and roots and tubers are classified under this head. This food group includes, several commonly used vegetables like brinjal, okra, french bean, guar bean, various gourds, tomato, etc. They not only add variety to the diet but also provide vitamin C and some minerals. These vegetables are also a source of dietary fibre in the diet and provide bulk to the diet.

5.7.5 Condiments and Spices

These are accessory foods mainly used for flavouring food preparations to improve palatability. These are used in small amounts and their contribution to nutrient intake is very limited. Some of the spices however are also rich in iron, trace metals and potassium. Some of the condiments like chillies and coriander may provide some beta-carotene. Green chillies also provide vitamin C. Most of the spices contain a high level of tannin which may interfere with iron absorption. Spices also contain several pharmacologically active substances like choline, biogenic amines etc. Some of them like garlic have antibacterial property and inhibit purifying bacteria.

To maintain good health and fibre intake, one should consume 3-5 servings (one serving is 80g) of vegetables per day. Daily recommendation by FAO/WHO is 450 grams/day. However, there has not been much improvement in consumption over the years in South Asia Establishment of home gardens on a large scale can help improve year-round vegetable production, ensuring nutritious vegetables are available to the smallholder farmers, poor households and other keen home gardeners

6

Seed Production and Storage

6.1 What is a Seed?

A seed is referred as a kernel in some plants. It is a small embryonic plant enclosed in a covering called seed- coat, usually with some stored food. It is the product of the ripened ovule after fertilization and some growth within the mother plant. The formation of the seed completes the process of reproduction in seed- plants, with the embryo developed from the zygote and the seed- coat from the integuments of the ovule. "Seed" term may also be used for a propagative part of a plant such as a tuber or a spore.

6.2 Why Produce Seeds in a Home Garden?

Main problem in vegetable cultivation is to obtain good quality seeds. And home gardeners normally do not know from where to acquire good seeds, and even if they do, it is difficult to procure seeds in small quantities. Seeds available in the local markets are not always of good quality, not available in small packs, and often do not germinate properly. To purchase seeds every year is very expensive. Therefore, availability of good quality seeds at the household level only would make it possible to grow vegetables successfully throughout the year. Production of seeds in home garden is cheap, seeds can be available at any time, and gardeners can be independent, and seed quality would be assured and there would be no difficulty in procuring small quantity of seeds. The seed germination, vigour and size (the three aspects of seed quality) may influence crop yield through both direct and indirect ways. Seed may account for 5-10% of crop production but seed quality can be responsible for 20-50 % crop yield. Before understanding seed production, we need to understand various classes of seeds.

6.3 Seed: Classes, Production Practices and Maintenance

Classes of Improved Seed: The certification programme classifies seed generations into the following.

6.3.1 Breeder Seed

It is the purest quality seed or the vegetative propagating material of a named variety produced by the breeder or the breeding institute, which released the variety. It is 100% genetically pure with all morphological characters that the breeder incorporated. To maintain purity, positive selection is carried out during every seed production cycle, and only a few of the best plants that are true- to- type of the variety are selected. Pollination is under strict control. Breeder's seed is used to produce foundation seed. In case of self-pollinated species, mass selection may be regularly practiced to retain genetic purity of the variety. Off-type plants are promptly eliminated, and care must be taken to avert out crossing or natural hybridization and mechanical mixtures.

6.3.2 Pre-basic Seed

Several generations of multiplication may be carried out as for breeder seed, to build- up population of basic seed. The generation between the breeder seed and the basic seed is the pre-basic seed. This step may be by-passed if a large quantity of breeder seed can be produced, stored, and released gradually as needed.

Basic/foundation seed: Foundation seed is the progeny of the breeder seed or proven basic seed by direct increase. It is generally pure and is the source of registered and/or certified or truthfully labelled seed. Production of foundation seed is the responsibility of seed corporations. It is produced on Government farms at the experiment stations; Agriculture Universities; and on cultivator's field under strict supervision of research scientists and experts.

6.3.3 Registered Seed/Certified/ Truthful

Label seed is produced from or is the progeny of the foundation seed; is genetically pure and is usually produced by progressive farmers according to technical advice and supervision provided by the experts. Often registered seed is omitted and certified seed/truthfully labelled seed is produced directly from the foundation seed. The seed produced from basic seed is known as certified seed, first generation (C_1). Seed produced from the latter is known as the certified seed, second generation (C_2), and so on. The certified seed is annually produced by progressive farmers according to the standard seed production practices. To be certified, the seed must meet certain rigid requirements regarding purity and quality. Certified seed is available for general distribution to farmers. This is generally produced by Seed Corporations and progressive farmers. Due to the number of generations for large-scale production, the variety may lose its genuineness. The number of regenerations of the certified seed should, therefore, be controlled.

Asexually propagated crops: Asexually propagated crops are maintained by asexual reproduction. Mechanical mixtures are avoided and off-type plants are promptly eliminated.

6.4 Requirements for Good Seed

Seed has to meet certain requirements before it will be certified for distribution as improved seed. Most important is the seed must be of an improved variety, released by the Variety Release Committee for general cultivation; this is essential for the seed to be certified. Other requirements are related to genetic purity, freedom from weeds, diseases and pests and quality of germination of seeds.

The home gardens need a small amount of seed per variety but of many varieties/ species. The demand for seed is year-round and the species and varieties required vary with season and location. Farmers usually save seeds from their own crops for the next planting. Some buy mature fruits from the local market and use the seeds for planting.

6.5 Seed Production and Processing

Seed is the basic input and a primary requisite for successful vegetable growing. The potential yields of the new high- yielding varieties cannot be realized unless the purity of the variety is maintained year after year. This can only be done by producing vegetable crop seeds – carrier of the genetic purity of the variety on the scientific lines. Production of seeds of high quality vegetable crops requires considerable technical skill and a number of rigid requirements must be fulfilled to ensure high purity and germination of seed. Seed multiplication involves two separate steps: (i) seed production and (ii) seed processing.

6.5.1 Seed Production

It requires improved cultural practices, efficient weed, disease and pest control, optimum irrigation and fertilizers inputs and some other specific operations. Additionally farmers experience is very much required to identify/select proper plants and fruits for harvesting seeds. Only progressive farmers are given the responsibility of seed production since they have all the required inputs for raising a good seed- crop.

6.5.2 Seed Processing

It is ordinarily done by different Seed Corporations, who have the facilities for large- scale processing of seed. But a gardner may develop his own facilities for seed processing if he is producing seeds on a large scale. Good quality seed can contribute to increased vegetable yield, as high as 30%.

The seed producer should be familiar with vegetable cultivars and their various characteristics, breeding system, climatic requirements, cultural operations, isolation distances, rouging practices, serious insect diseases, weeds, drying ,processing methods etc.

Main Considerations for Producing Seed in a Home Garden: Seeds should be saved from healthy, vigorous plants only. The plants from which seeds are to be saved should be marked and observed during the entire growth period. No seeds should be extracted from diseased and/or pest-infected plants. Undesirable plants should be removed.

If the large amounts of seed are required, a plot should be reserved for seed production alone; in isolation from other similar crops.

6.6 General Aspects of Seed Production

6.6.1 Selection of a Right Variety

A vegetable variety to be grown for seed production must be genetically pure and adapted to photoperiod and temperature prevailing in the production areas. The choice of a right variety is the first step for successful seed production programme, e.g. A short- day onion variety and annual type cabbage produce abundant seeds in winter season; and long- day onion variety and biennial cabbage type do not produce flowers during the winter season, because of long day requirement of this onion variety for its growth and development and vernalization requirement in cabbage variety flowering.

6.6.2 Selection of Suitable Areas for Seed Production

Areas selected for seed production should be under the agroclimatic zones with moderate rainfall (30-40 inches or 75-100 cm), humidity (40-50% RH), temperature (15-20⁰C), and gentle winds. These conditions favour more flowering and pollination in most of the vegetables.

a) Isolation Distance

Genetic purity of the seeds is another important character of good quality seeds. This can be maintained year after year if adequate isolation distance is provided to the seed- crop. In the self-pollinated crops, an isolation distance of 25 to 50 m may suffice to avoid mechanical mixture at the time of planting or harvest. Cross-pollinated crops require an isolation of 400 to 1,600 m depending upon the amount and type of out-crossing.

Table 6.1: Isolation Distance Requirement for Different Vegetable Crops

Crops	Mode of Pollination	Viability (yr)	Approximate Isolation Requirement (metres)		Seed Count
			Foundation Seed	Certified Seed	
Solanaceous crops					
1.Tomato	SP	4	50	25	400/g
2. Brinjal	SP	4	200	100	225/g
3.Chillies & Capsicum	SP	2	400	200	160/g
Cole crops					
1. Cauliflower	CP	4			315/g
2. Cabbage	CP	4	1600	1000	315/g
3.Knol khol	CP	3			315/g
Root crops					
1. Beet	CP	4	1600	1000	55/g
2. Carrot	CP	3	1000	800	800/g
3. Radish	CP	5	1600	1000	90/g
4. Turnip	CP	4	1600	1000	525/g
Legumes					
1. Frenchbean	SP	1	20	10	4/g
2. Cowpea	SP	3	20	10	8/g
3. Peas	SP	3	20	10	5/g
4. Cluster	SP	2	20	10	
5. Dolichos- bean	SP	1	50	25	
Cucurbits					
1. Muskmelon	CP	5			45/g
2. Pumpkin	CP	4			7/g
3. Watermelon	CP	4			7/g
4. Sponge gourd	CP	2	800	400	
5. Ridge gourd	CP	2			9/g
6. Summer squash	CP	4			40/g

Contd...

Crops	Mode of Pollination	Viability (yr)	Approximate Isolation Requirement (metres)		Seed Count
			Foundation Seed	Certified Seed	
7. Winter squash	CP	4			
8. Cucumber	CP	5			
Leafy vegetables and salads					
1. Lettuce	SP	6	50	25	900/g
2. Amaranthus	CP	3	400	200	
3. Spinach	CP	3	1600	1000	100/g
Others					
1. Onion	CP	1	1000	400	300/g
2. Okra	SP	2	400	200	18/g

b) Steps for quality seed production

The following are steps for quality seed production.

i)   Land Requirement: The land should be levelled, fertile and free from noxious weeds, common to the crop. In some cases, the field to be used for seed production should not have been used for growing same crop in the previous year.

ii)   Plant Selection: The time for maintenance and selection of a variety needs to be carefully identified. The best plants in terms of growth and yield and the best fruits should be selected for seed production. Any plant or fruit with suspected symptoms of disease and pest attack should be excluded.

iii)   Cultural Practices: Recommended cultural practices must be followed for raising a good seed- crop. Recommended doses of fertilizers and irrigation water must be applied for high yields of high quality seeds. Poor cultural practices would give lower yields and seed of smaller size, which would be rejected at the time of grading. This would drastically reduce profits of seed growers.

iv)   Plant Protection. Adequate measures must be taken to protect seed- crop from diseases and pests. Insect- pests and diseases may cause considerable damage to crop, reducing yield and seed quality. This rigid requirement with respect to diseases is prescribed to prevent epidemics owing to contaminated seeds.

v)   Weed Control: Effective weed control is a must for good seed production. Weeds reduce crop yield and weed seeds contaminate seeds. Certain weeds are classified as objectionable weeds by the seed certification agency ;so the seed field is generally required to be free from such weeds.

vi) **Bagging:** When only a small amount of seeds is needed, cover unopened flowers with a paper bag. This is applicable for crops with a high but not 100% self-pollination such as pepper and eggplant. Flowers of cucurbits can also be bagged. In this case, both male and female flowers should be bagged, but hand-pollination is required.

vii) **Caging:** Cages can be used for vegetables that flower over a long time and to prevent insects from transmitting pollen from two nearby varieties of the same crop. Bamboo/iron rods can be stuck onto ground to make an arched tunnel covered with nylon mesh. There is needed to hand pollinate plant to ensure seed- set, or introduce bees into the cage if they are cross-pollinated species. Advantage: all the fruits which set, while the cage is closed, will be pure seed.

viii) **Rouging:** It is removal of plants which are off-types e.g., phenotypically different from the plants of the variety under certification. It is an important aspect of seed production and is necessary to avoid out -crossing and mechanical mixture. Off-type plants are regularly removed from the field either by uprooting or by cutting at the ground level. These plants may differ in plant height, leaf characters, flowering time, maturity etc.

6.6.3 Types of Seeds Required

The following are seed types.

i) **Hybrid Seeds:** These are purchased and are expensive. They require significant inputs to produce a superior yield.

ii) **Open-Pollinated Seeds (OP Seeds):** These are satisfactory for the purpose of home garden seed production. It is easy for home gardeners to maintain quality of seeds for self-pollinated vegetables. There is challenge for cross-pollinated vegetables, which tend to become modified and diversified.

6.6.4 Pollination:

For successful seed production, it is important to understand pollination behaviour of each vegetable. There are three ways in which vegetables are pollinated: self-pollination, cross-pollination and often/partial cross-pollination

a) Self-Pollinated Vegetables (Beans, Tomato, Lettuce)

They are complete with both male and female parts. Pollen of one flower fertilizes the same flower or another flower on the same plant.

b) Cross-Pollination

It occurs when an insect or wind transfers pollen from the flower of one plant to the flower of another (pumpkin, radish, batishak)

c) Partial Cross-Pollination (Okra and Eggplant)

When flower morphology of self-pollinated vegetables result in a certain amount of cross-pollination.

Table 6.2: Mode of Pollination in Vegetable Crops

Vegetable	Mode of Pollination	Pollination Medium
Red/green amaranth	Cross	Wind
Radish, carrot, beet, turnip	Cross	Insect
Cabbage, cauliflower, broccoli, kohlrabi, batishak, china shak	Cross	Insect
Pumpkin, ash gourd, bottle gourd, snake gourd, bitter gourd, ridge gourd, cucumber, melon, watermelon	Cross	Insect
Tomato, capsicum	Self	Insect
Okra, eggplant	Partial Cross	Insect
Kangkong	Cross	Insect
Spinach	Cross	Insect/Wind
Basella	Cross	Wind
Beans, french bean, yard long bean, soy bean, dolichos, cowpea	Self	Insect
Lettuce	Self	Insect

6.7 Seed Collection and Extraction

6.7.1 Guidelines for Better Seed Collection

Harvest the selected plant at the proper stage of maturity depending on the type. Extract seed from fruit/pod using appropriate techniques: wet seed extraction by fermentation for seeds contained in fleshy fruits; and threshing after drying for seeds contained in pods or cobs. Collect seeds in the dry season rather than the wet season. This avoids disease problems. Collect seeds on a dry, sunny day if possible. Extract seeds in a clean and empty area to avoid accidental inclusion of non-selected plants. Separate seeds which have been collected for seed production from those collected for vegetable crop production

a) Harvesting of Seeds

The seed crops of vegetables can be grouped according to the state of the seed:

Seeds of fleshy fruits: There are two types of seeds under this group: Seeds with mucilaginous coating , e.g. tomato and cucumber; Seeds without mucilaginous layer, e.g. pepper, eggplant, melon and squash

a) Seed Drying on the Plant

At physiological maturity, when the development of the embryo and accumulation of storage material are almost complete, moisture content of the seed is about 50% wet basis. Then seeds start losing moisture while they are still attached to the mother plant because of the sun and the wind. Sooner they are harvested and then dried by natural or artificial means.

Harvesting dry vegetable seeds: Beans and other leguminous vegetables are usually allowed to dry on the vines and harvested before shattering of the pods start. Likewise, pods of *Brassica* crops are harvested when 70% of inflorescences are ripe (turn brown or black) and before pods shattering. Fruits of *Luffa* species are also left on the plants to dry. Other vegetables with seeds harvested dry are okra, onion, kangkong, sweet- corn, lettuce and carrot.

Table 6.3: Maturity Indices of Some Vegetables for Seed Production

Vegetables	Maturity Indices
Garden bean	Pods mature and yellow
Cowpea	Two-thirds of pods turn brown
Dolichos bean	Pods dry and yellow
Garden pea	Seeds fully developed and hard red ripe
Pepper	Ripe or ripening
Tomato	Beyond edible stage
Eggplant	White fluff (30-50%) on heads
Lettuce	Seeds dark brown in colour
Cabbage	Pods dark brown in colour
Cauliflower	Pods turning brown
Watermelon	Edible maturity
Cucumber	Fruit pale yellow/golden

Table 6.4: Important Features in Seed Production of Vegetable Crops

Tomato	Good sized, well ripe fruits free from diseases are selected
Brinjal	Harvest fruits when they are greenish yellow or brown. Floating seeds should be rejected. Dry seeds in partial shade immediately
Chilli & Capsicum	Sweet pepper (*Capsicum*) readily crosses with hot chilli. Both types are not grown for seed production at the same time at the same farm. Harvest at the red ripe stage and dry
Radish	For quality seed, roots are to be pulled out, examined and the best root then replanted;cut 2/3 top and ¼ root
Turnip	It crosses readily with mustard; proper isolation has to be given
Okra	Select pods after 3 nodes
Amaranthus	Seed crop is harvested when most of the leaves turn yellow
Cucurbits	The following crops of this group should not be grown together for seed production; Muskmelon – long melon, watermelon, round melon, sponge gourd – ridge gourd
	Bottle gourd, sponge gourd and ridge gourd – extract seeds from dried fruits
	Pumpkin, cucumber, bitter gourd, ash gourd – extract seed from ripe fruit

i)   Pods/Fruits Drying : In tropical countries with wet weather, fruit- drying on the plant itself is not advisable. The physiologically mature fruits must be harvested and air-dried immediately under shade or under sun, whenever possible. Hanging wet fruits/pods above the wood stove can help hasten drying and may prevent insect-pest infestation. Sometimes it may be necessary to split wet, mature, green legume pods to remove seeds to dry them faster. In this case, however, the seeds should be handled carefully because they are susceptible to damage by bruising. When drying with artificial heat, the temperature must not be above 30°C when seeds still have a high moisture content, greater than 20% (usually after harvest) to avoid heat injury. The drying temperature can be increased later in to 35°C -40° C.

ii) Threshing and Processing of Dry Pods/Fruits: As the volume is small, dry pods/fruits (Fig.1) can be threshed manually after harvesting by beating with a stick or rubbing and splitting by hand. Threshing is easier and more efficient if the harvested pods/fruits are first dried directly in the sun to make them more brittle. The clean seeds are further dried to about 8% moisture content (MC) directly in the sun on a tray, drying mat, or any suitable container. The seeds should be spread out in a thin layer to hasten drying. The drying place must be well-ventilated, and the seeds are turned from time to time. Drying time will depend on the moisture content of the seeds which can be determined by its hardness/brittleness. When beaten or crushed with a stone, a dry seed of about 8-10% moisture content is hard and brittle. The drier the seed, the harder and more brittle it becomes; cracks or shatters with a characteristic popping sound. Wet seed is elastic and therefore will not crack/break.

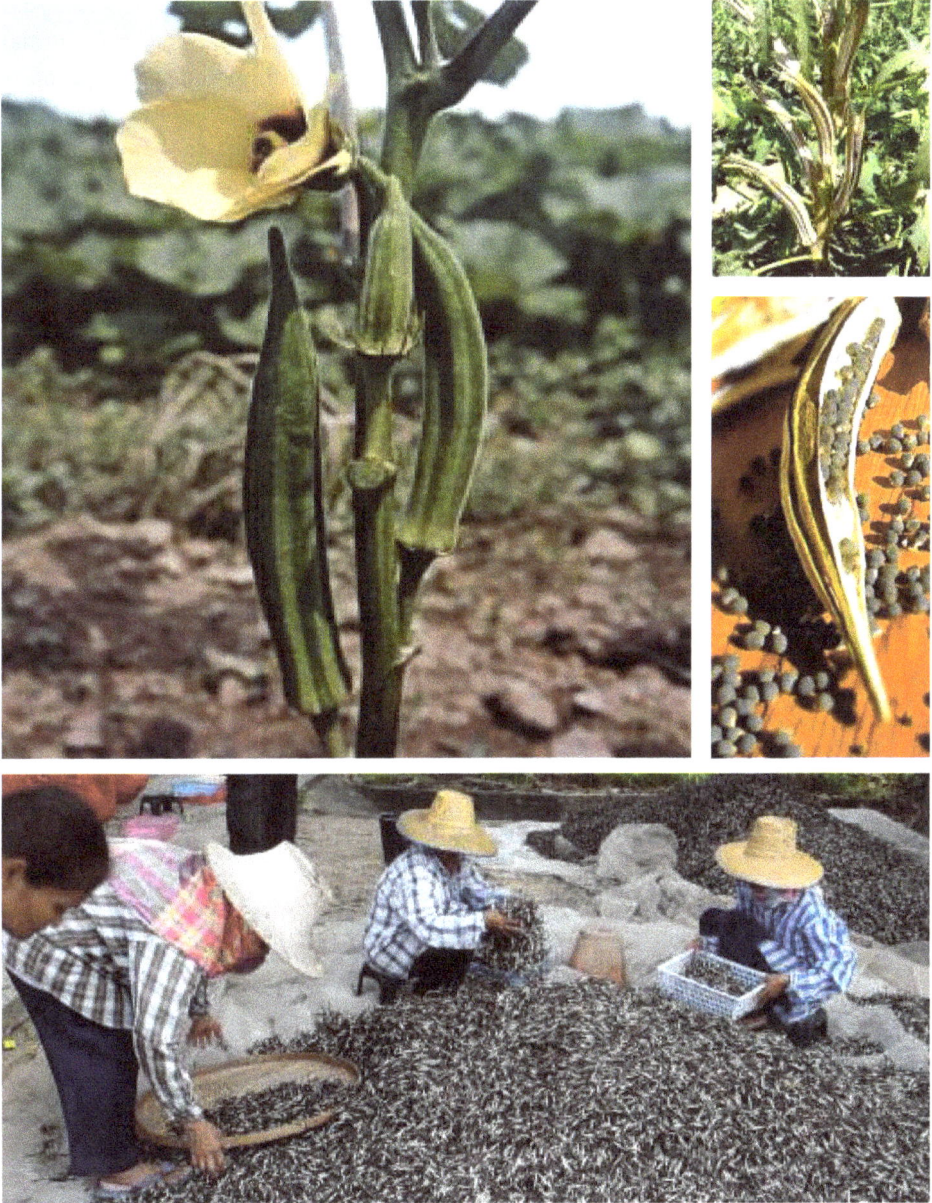

Fig. 6.1

iii) Extracting and Processing Seeds of Fleshy Fruits: Seeds with mucilaginous: (Tomato): Seeds of the fully ripe fruits can be extracted right after harvest. The fruits are cut and crushed in a plastic bucket or wooden container. The crushed fruits in the form of slurry are left to ferment in the shade away from direct sunlight and rain. The pulp should be stirred regularly to help release seeds and to prevent mould growth. Fermentation is complete when bubling activity (CO_2 released due to fermentation) is reduced, the slurry cools down and decreases to the initial volume, and the supernatant of the slurry clears up. A more practical indication is the sinking of seeds to the bottom of the container. Natural fermentation takes one to two days at a pulp temperature of 22 – 27 °C, and two to four days at 15-22 °C. If fermentation is sufficient, no mucilaginous matter is left on the seeds and the floating pulp can be easily separated. The seeds are then washed and cleaned, dried in the same way as seeds of dry fruits.(Fig. 6.2)

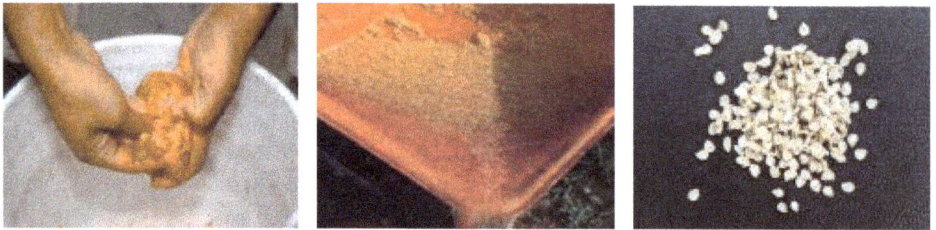

Fig. 6.2

Seeds Without Mucilaginous Layer: (Sweet pepper**):** Seeds are extracted directly without fermentation. Extraction is done by splitting fruit lengthwise and removing seeds manually with a suitable tool. The fruit wall is separated from the core of the fruit and the seeds are rubbed manually or washed out by rubbing in ample water. Clean, wet seeds are then spread out to dry.

6.8 Seed Extraction in Various Vegetables

6.8.1 Amaranth

The seed-crop is harvested when most of the leaves turn yellow. The bundles of seed bearing shoots are dried in the sun for a few days. For actual seed extraction, these bundles are beaten and seeds are collected. It is also possible to have second harvest of seeds from the same crop.

Fig.6.3

6.8.2 Beet Root

The so called beet root seed in reality is a fruit. Each fruit contains 2 - 6 seeds. The actual seeds are small. Kidney- shaped and brown. Other practices are similar to turnip.

6.8.3 Brinjal

Fruits are harvested at full ripe stage when colour changes to bright yellowish. Seeds are extracted by simple mechanical crushing. Crushed fruits are mixed with water. Sedimented seeds are separated and dried to 8% moisture level. Recovery of brinjal seeds is always higher from medium- sized fruits, and quality seeds are recovered in about eight pickings.

Acid method: Fruits are pulped in an electric pulper and treated with commercial HCl at 1:40 ratio (25 cc/kg of pulped fruit), stirred for 25 to 39 minutes and washed well and dried to 8 to 10% moisture content. A maximum seed recovery of 5% to the total weight of fruits was registered by acid method. The average seed yield ranges from 360 to 400 kg/ha.

6.8.4 Carrot

Only the tropical varieties produce seeds in plains; temperate varieties produce seeds in hills only. More quantity of seed is produced when seeds are left in situ . For quality- seed production, roots should be dug out, selected and then transplanted. Cut 2/3 from the top and ½ of the root while transplanting. Carrots spiny little seeds form an umbel on the top, which eventually dries and seeds are ready to be harvested. Individual flower produces a pair of seeds

6.8.5 Cauliflower and Cabbage

Seeds of early and mid season varieties are produced in plains while seeds of late cauliflower are produced in hills. Seeds are produced through in-situ method with scooping. The best and most economical method is to leave cauliflower heads in situ. Transplanting or scooping reduces yields. Cauliflower, cabbage and knolkhol cross one another; therefore suitable isolation should be provided for them.

Very rarely cabbage produces seed in plains; the seeds are raised only in hills. Seeds are raised by two methods: Head to Seed; Seed to Seed. Head to seed method is employed in raising nucleus seeds. In seed to seed method, higher yields are possible. Quality in both the cases is same.

6.8.6 Chillies and Capsicum

The crop starts yielding green chillies 2 months after transplanting and dry chillies, 90 days after transplanting. The crop lasts for 5-6 months after transplanting, depending upon the duration of the variety; fruits are harvested at fully ripe red stage. Four to five pickings of green chillies or two to three picking of red chillies are done (Fig. 6.4). All the virus infected plants should be rouged; diseased fruits should be called out to keep away seed- borne diseases. Seed can be extracted from fully dried chilli pods (sun dried) mechanically. Seeds after extraction could be used immediately for sowing as they have no dormancy. They are extracted from fresh fruits or from dried fruits; 8% moisture is retained before storage. Storing of dried fruits as such prolongs seed viability much longer than stored extracted seeds. Recovery of seeds from 5 to 6 cm long fruits was maximum ranging from 51% to 68% in different pickings. Seeds extracted from fruits dried under sun scored 11% increased germination over fruits dried under shade. Sweet pepper (Capsicum) crosses with hot chilli, it is, therefore, necessary to ensure that both types are not grown for seed production at the same time and at the same farm. Harvest fruit at red ripe stage and dry. Extract seeds by breaking dry shells.

Fig. 6.4

6.8.7 Cucurbits

The following groups of crops should not be grown together.

- Snap melon – long melon – muskmelon
- Watermelon – round melon
- Sponge gourd – ridge gourd
- *Cucurbita moschata – C. maxima*

Seeds of bottle gourd, sponge gourd and ridge gourd are extracted from dried fruits,and of pumpkins, cucumber, bitter gourd, ash gourd are extracted from ripe fruits. Anthesis takes place during morning hours (4 AM to 9 AM) in majority of the cucurbitaceous crops except bottle gourd, sponge gourd and coccinia; in which anthesis takes place during evening hours. Since cucurbits are insect pollinated, bees are essential for good yields. Inadequate pollination results in poor fruit shape and excessive blossom drop. Hence, it is advisable to keep 3 bee -colonies per hectare during flowering time for a successful crop. Roughing of off-type plants should be done at 3 stages : a) before flowering; b) at flowering and fruiting; and c) at fruit maturity. In fleshy fruited vegetables, fermentation method or acid method is followed to extract seeds. Fermentation of pumpkin and squash seeds should be avoided. In case of dry fruits, seeds are extracted by breaking open fruits by making hole at the blossom- end of the fruit. After washing/cleaning, the seeds should be dried under sun or in seed driers. Drying temperature should not exceed 38-40°C. Moisture content should be reduced to 6-8% for safe storage.

6.8.8 Fenugreek

Remove unhealthy less vigorous plants. Harvest pods when majority are yellow, some dry-cure-thresh.

6.8.9 French Bean

For seed production, it is preferably grown at slightly higher altitudes, and at places where rainfall is comparatively low. It is important to rogue seed crop carefully to maintain genetic purity. Harvesting is done usually by hand when a large percentage of pods are fully ripe and the remaining have turned yellow. Harvesting should be started before the lower pods become dry enough to shatter. The crop after harvesting is left in the field to dry for about seven to ten days. Later, it is threshed by bullocks or by threshers. While harvesting, care should be taken to keep mechanical injury to minimum. After threshing and cleaning, the seed should be dried to moisture content below 9% before storage.

6.8.10 Kangkong

When seed pods are mature, uproot plants when dry weather is assumed to be for several days. Uproot plants and keep them in the field for a few days. The plant spread will curl into a loose bundle. Turn bundle several times to dry uniformly. After 3-4 days of drying, thresh plants. Seed can be cleaned by winnowing.

Fig. 6.5

6.8.11 Okra

The dry pods should be harvested for seeds before shattering. Seeds are threshed out from pods by beating with sticks or by passing through rollers. The seeds are cleaned and dried to a moisture content of 8-10%.

6.8.12 Onion

It is a biennial plant, requiring 2 seasons to complete its life- cycle. In the first year, the onion seedlings yield bulbs. In the second, bulbs when planted at optimum period produce seeds. Onion- seed is known for its poor keeping quality, and after one season it deteriorates fast. Two methods are employed for seed production:Most commonly is bulb to seed method involving 2 seasons; Smaller volume of seed is produced by seed to seed method.

Though a good amount of seeds can be obtained in the first year if some of the plants are allowed to flower, yet this is not a proper method. But in such cases: selection is not possible, and yield is low ; such seeds produce thick- necked bulbs inferior in quality. Seed- producing area should have low humidity, mild cold temperature during initial growth, followed by increase in temperature at later stages of crop. While onion is in flower, clear, bright days are necessary to ensure insect activity for pollination. In the common method of seed production, onion- bulbs are harvested, stored for sometime and then replanted in the proper season for seed production. Mother bulbs are grown in the same way as the crop for market and are stored in well ventilated storehouses. Generally medium sized bulbs are selected for planting for seed. About 1,500 kg of bulbs are required for planting for seed. The spacing depends on the size of bulbs. However, the ideal spacing is 50 cm between rows and 20 cm within rows. All the seed heads do not mature simultaneously; there is usually only one cutting which is made sufficiently early when black seeds are visible to avoid seed shattering . The seeds are threshed out from seed heads when they are fully dry by beating with sticks.

6.8.13 Pea

Pea crop is known for producing off- type plants, which rapidly multiply if not rogued every year. It should be harvested when 25% of the pods are ripe. Cut vines and keep for full drying.

6.8.14 Radish

Radish pods turn brown seamless or indehiscent when they are ready for harvesting. Rub them between palms to release seeds. Winnow out lighter pieces of stem or pod ,which are introduced along with the seed.

a) Cleaning and Grading of Seeds

It is done to remove undesirable materials from desired ones. The undesirable materials include: inert matter, weed seeds, other crop seeds, light-damaged or deteriorated seeds. Home gardeners perform this process mainly by winnowing and by gravity separation using bamboo winnowing tray.

b) Seed Drying

It increases longevity, viability of the seed and facilitates storing. Drying removes excess moisture ,which causes excessive respiration, heating and fungal invasion. The amount of drying required depends on air, temperature, relative humidity and wind velocity.

c) Points to Remember

Choose a spot which is sunny all day, spread seeds thinly over a mat and not on the ground; seeds should be turned over gently 4-5 times a day to ensure their even drying , in the evening bring the seeds inside, continue the process until they are properly dried.

d) Methods to Test if Seeds Dried Enough for Storage

Large, thin seeds would break with a snapping sound when twisted between fingers, e.g. pumpkin and gourds. Large, thick seeds which cannot be broken between fingers, would break with a cracking sound when bitten between front teeth,e.g. beans, soybean, maize. Small seeds would break with a cracking sound when squeezed between fingernails, e.g. amaranth, radish

e) Homestead Seed Packaging and Storage

When seeds are dried and cleaned, they have to be stored properly before planting. The seeds are placed in an appropriate package or container and stored until planting to maintain high viability.

Seed moisture content and storage temperature are two important factors ,which affect seed viability in storage.

6.8.15 Spinach

The seeds are usually produced on hills but those of smooth leaf varieties are produced in plains. It produces seed in 150 days. Seeds are formed and mature in an indeterminate growth pattern, beginning on older, lower branches and continuing up to the flower stalk through the season. In most cases, only about 75% of the seeds on any given plant would reach maturity at harvest. One method of gauging the maturity of a spinach seed -crop is to make a visual assessment of the percentage of seed on most plants that have turned tan-brown, typical of mature spinach seed; harvest the crop when 60 to 80% of the seeds have attained the colour.

6.8.16 Tomato

Good sized, well ripe fruits free from disease are selected and about 200 kg fruits are required to produce 1 kg seeds.

a) Seed Extraction Techniques

The seed can be extracted by two method.

i) Fermentation Method: Red- ripe fruits are crushed to have juice and seeds in a container. Leave them to ferment for 24-48 hours. Then seeds are washed thoroughly to remove mucilaginous material around them and then they are dried. Seed moisture for good storage should be around 8%. This by far is the best method for seed extraction.

ii) Acid Treatment: The seed pulp extract is treated with 50% hydrochloric acid in 3:1 ratio (3 parts juice and 1 part hydrochloric acid) for 1 hour. The content is frequently stirred. Normal washing in cold water is followed and then seeds are sun-dried.

b) Alkali Method

Sodium hydroxide or washing soda at the rate of 500ml in 4 litres (1:8) of water is mixed with equal quantity of pulp. Keep it for one night (10-12 hours) and wash and dry.

c) Mechanical

De-seeding machine is used. Seed and juice are separated from pulp. seed is separated and dried to 8% moisture before storage. Seed quality assessments indicated increased vigour and germination potential of seed extracted from large and medium fruits over small fruits.

d) Seed Production in Hybrids

The male parent should be sown and planted at least 15 days prior to the female parent, to ensure availability of abundant pollen at the time of crossing. Buds, one or two days prior to anthesis, should be emasculated by holding bases of two petals together with a pair of sharp forceps. The petals come out along with anther cone. Care should be taken not to injure style, stigma or ovary. After pollination, one or two sepals of these buds should be clipped by hand for easy recognition of crossed fruits at harvest. In case, the female parent is staked, one can obtain about 20 crossed fruits per plant. The seed yield can be between 40 and 75 kg per hectare, depending on crop condition, efficiency of crossing, seed content of the seed parent and temperature at pollination. Pollen can be collected in abundance by sun- drying freshly opened flowers(bright yellow anthers with greenish tops, petals fully open)for a day and then sieving them through a 2- layered muslin- cloth. The pollen can be easily stored under anhydrous conditions and cool temperature of 10°-12 °C for a week.

For maintenance of seed quality, the seed producers should strictly adhere to the following.

- Have through knowledge of the variety and its chief attribute.
- Be ruthless in roguing out off- types. Roguing should be carried out at least twice, once before fruit maturity and once afterwards.
- Maintain a rigid schedule of crop protection measures to ensure a healthy crop.
- Remove all seed- borne virus affected plants – e.g., Tobacco mosaic virus infection in tomato.

6.8.17 Turnip

It crosses readily with mustard, so proper isolation has to be given against contamination. The biennial temperate types produce seeds only in hills, while annual Asiatic types can produce seed in plains. In-situ crop gives better yield than transplanted one. There is no difference in quality of seeds. In transplanting, selected roots are pruned and tops clipped leaving the crown intact.

6.9 Factors Affecting Seed Storage

a) Humidity

Seeds absorb moisture from storage environment. High humidity levels cause increased seed respiration rate and use stored energy. Seeds should be dry enough (seed moisture content around 7-8%) before storage, and kept in air-tight container such as screw-top jar.

b) Darkness

Exposure to sunlight shortens life of seeds. Use dark-colour jars or non-transparent containers to protect seeds from sunlight. If clear jars, place them in paper bags to shield out sunlight.

c) Temperature

For most vegetable seeds, temperature below 15ºC is ideal. Seeds in an air-tight container can be placed in refrigerator. For short-term storage, keep seeds in a cool and shady dry place.

d) Storage Containers

Preserving seeds in a suitable container(Fig. 6.4) would prevent direct contact of seeds with environment, and this is another approach for retaining viability. Different containers used include: Paper and cloth bags – for short term storage; metal, polythene, glass, laminated and foil containers for moisture- proof and long- term storage.

If small quantities of many species and varieties are to be stored, a home-made desiccator can be used. It can be made from any large-mouth glass jar with a good rubber-gasket top, (e.g., food preservation bottle, or small airtight metal biscuit tin with a good lid).

Fig. 6.6

Among the desiccants that can be used are silica gel, calcium chloride, quick lime, cooled-down fresh wood ash, and dry charcoal. Silica gel and calcium chloride can be reactivated for reuse by heating. If blue silica gel (silica gel with cobalt chloride as color indicator) is available, it can be used as an indicator of leakage when a transparent bottle or plastic bag container is used. Dry silica gel is blue and is pink. A small perforated polyethylene bag of this type of silica gel can be packed with seeds.

At the bottom of the jar, a layer of desiccant (e.g., silica gel), which is about 10% of the container space, is put. This is followed by a porous layer of paper, wire-mesh, or perforated metal plate as base for seed packages to stand on.

The seed packages should not be airtight; there will be free exchange of moisture between seeds and desiccant. Suitable materials for packaging in this case are paper, cotton and nylon netting.

Plastic bags should not be hermetically sealed; the moisture from the seed needs to be diffused out and be absorbed by the desiccant. Seed should be well dried at the time of storage and container should be sealed properly so that standards of germination could be maintained.

Table 6.5: Standards of Minimum Germination for Vegetable Seeds

Vegetable	Germination %	Vegetable	Germination %
1. Alliums		**4. Leguminous Crops**	70
Leek	70		
Onion	45	Broad Bean	75
Welsh onion	75	Pea	75
		Snap (French) Bean	75
		Yard-Long Bean	75
2. Crucifers		**5. Solanaceous Crops**	
Broccoli	75	Eggplant	65
Cauliflower	75	Pepper (Sweet, Hot)	65
Chinese cabbage (Heading)	75	Tomato	75
Chinese cabbage (Non Heading)		**6. Other Tropical Vegetables**	
Common cabbage	75	Amaranth	70
Kale	75	Kangkong	70
Mustard	75	Okra	80
Radish	75	Sweet Corn	75
Turnip	80	**7. Other Cool-Season Vegetables**	
3. Cucurbits		Asparagus	75
Bitter Gourd	65	Carrot	70
Bottle Gourd	80	Celery	65
Cucumber	80	Lettuce	70
Melon	80	Spinach	60
Luffa gourd	75		
Pumpkin	80		
Squash	80		
Watermelon	80		
Watermelon (Seedless)	60		
Wax Gourd	70		

6.10 Principles for Storage

Bottles, tins and glass jars with tightly fitting lids or stoppers can be used. Container should be quite dry before putting seeds. Do not use paper, leaves or grass to plug a bottle as these would allow moisture to enter the bottle. Use rat and mice-proof containers. Different varieties of seeds can be put into separate plastic bags and then into large tins or glass jars. Toasted rice or ash can also be kept inside the container to draw moisture out of the air and keep seeds dry. Do not put small amount of seeds in a large container. Heat kills seeds, so store them in a cool, shady and dry place. Do not place seeds directly over cooking place or in direct sunlight

6.11 Seed Quality Testing

Quality of the seed is a major factor which determines economic success of a crop. Poor quality seeds give a few and uneven plant stands from the very start, and they may be source of inoculum for some seed-borne diseases.True botanical seed can be classified into following two classes.

- Orthodox seeds are those which can be dried to low moisture level for storage at low temperature, e.g. most vegetables.

- Recalcitrant seeds are those which do not survive drying and freezing,e.g. chayote (*Sechium edule*)

6.11.1 Characteristics of High- Quality Seeds

High analytical purity: The seeds must be free from seeds of other species and varieties, weed seeds, and other impurities. Pure seeds may be separated from impurities during seed processing. Varietal mixture is more difficult to distinguish; this can be minimized by using superior stock and preventing crossing between varieties.

Good vigour and germination capacity: The seed is at its peak vigour and germination level at maturity, usually just before harvest, after which it starts declining. It is greatly affected by the following.

a) Crop Growing

Improper growing conditions, such as inadequate water and nutrients, result in poor development of endosperm. This gives rise to light or shrivelled seeds, which germinate poorly.

b) Seed Maturity

Immature seeds have less germination ability and vigour than mature ones due to limited food reserves in the immature seeds. Immature seeds may germinate well after harvest but lose viability after a few months in storage when limited food reserves may be exhausted. Immature seeds are also unable to absorb water and oxygen, which are important for germination.

c) Moisture and Temperature

The higher the moisture content and temperature during storage, faster is the respiration process, consequently, faster is the loss of viability of seeds. For each 1% drop of moisture content and for each drop of 5.6°C, there is doubling of lifetime of the seed.

d) Seed Damage

Seed injury may affect embryo, food storage, or the pathway of food to the embryo. It also makes the seeds susceptible to attack by microorganisms.

7

Pest Management

Diseases, insects and pests management is one of the greatest challenges to home gardener. Home garden pests are not very much different from field pests but the incidence as well as occurrence of these depends upon the environment and microclimate of a particular location. Insect- pests include enormous variety from tiny thrips, nearly invisible to naked eye, to large larvae of tomato hornworm. Diseases are caused by fungi, bacteria, viruses, and other organisms. Home garden vegetables show susceptibility to many insect and disease problems if varieties are not resistant. These problems are to be effectively controlled for good quantity and quality of vegetables.

7.1 Identification of the Problem

Many plant problems are not caused by insects or diseases but are related to temperature extremes, waterlogging, drought, damage caused by lawn mowers or overuse of chemicals. Poor nutrition can also make plants susceptible to pests and diseases. It is, therefore, necessary to carefully identify the problem before applying any control practices. Some insect damage may appear to be a disease, especially if no visible insects are seen. Nutrient problems may also mimic diseases. Herbicide damage resulting from misapplication of chemicals also can be mistaken for the problem.

In home gardens, damage to plants is more likely to be caused by environmental conditions such as drought, nutrient deficiency, sunscald, frost, salt or wind burn, , mechanical damage, than from pests. If a pest population causes a problem, then it should be identified (to species if possible). When pest problems are identified based on plant damage, careful identification is essential because similar looking damages can have different causes. For example, curled leaves from a viral disease looks much like damage caused by sucking insects. Spraying a pesticide would be useless if plant symptoms were caused by a disease. Also, damage caused by nutrient deficiencies or poor growing conditions may be mistaken for signs of presence of pests.

Identify pests by comparing specimens with pictures in reference books/ literature or on the internet; recognizing characteristics of damage or of the experiment and castings left by pests; consult experts for assistance

7.2 Diseases of Home Garden

There are several diseases and insect-pests of vegetable crops which also occur in home garden. Insect-pests cause several types of injuries and infarctions depending upon the causal organism and nature of the host- plant.

7.2.1 Diseases common in home gardens

iii) Damping off: It is caused by different pathogens like *Pythium aphanidermatum, Fusarium* spp., *Phytophthora* spp., *Rhizoctonia solani* etc. Its main stages are: **pre-emergence damping-off**, which results in seed and seedling rot before they emerge out of the soil; **post-emergence damping-off**, which causes infection of young, juvenile tissues of the collar at the ground level. Plants/seedlings show gradual wilting and ultimately damping off. Internal discolouration of vascular tissue occurs with no root damage. Collar portion of the plants rots and ultimately seedlings collapse and die. Infected tissues become soft and water soaked.

iv)   Root- rot / crown- rot: It is caused by *Fusarium* spp., *Rhizoctonia* spp. , *Macrophomina* spp., *Verticillium* spp. *Sclerotium rolfsii, Sclerotinia sclerotiorum* etc. Infection occurs at the root level or at the stem base, adjacent to soil. The plant shows gradual wilting and complete death within 3-4 days. Uprooted plants show root rotting. In hot- wet season, plant shows severe wilting ,and dies suddenly.

Wilt: It is caused by early or late infection by mostly soil -borne pathogens, including *Fusarium* spp., *Rhizoctonia* spp. *Macrophomina* spp. *Verticillium* spp. *Xanthomonas*

spp. *Pseudomonas* spp. *Ralstonia* spp. *Pythium aphanidermatum, Fusarium* spp., *Phytophthora* spp., depending upon the host- plant. Plant shows wilting of lower leaves during hot daytime and recovers at night. Initially plants show temporary wilting symptoms, which become permanent and progressive. Within a few days, a sudden and permanent wilting occurs. Fungal pathogen affected leaves of the plants show yellowing, loose turgidity and drooping symptoms. In case of bacterial wilt, plant does not show yellowing with wilting symptoms. In older plants, leaves wilt suddenly and vascular bundles in collar region become yellow or brown. Eventually, the plant dies.

If a diseased stem is split length-wise, the vascular bundles appear as dark streaks or as brown colouring. In case of bacterial disease infection if the cut portion of root or crown region of stem is dipped into the water, clear bacterial oozing can be seen.

Fig. 7.1: Bacterial Wilt of Tomato

v)   **Leaf spot:** It is caused by fungal, bacterial and viral pathogens. The leaf spot size, shape, colour on both the surfaces differs as per the pathogen and respective plant species and or host- plant genetic material. Generally, fungal pathogens cause leaf spots of varying sizes and colours but the main . In bacterial leaf spot, only symptoms are observed but in fungal infection both symptom and signs are observed. Viral pathogens generally cause yellowing, curling, puckering, vein bending, vein clearance, stunting, etc., which suddenly enlarge. *Xanthomonas* spp. causes small minute water- soaked spots on the leaf lamina or margin which coalesce to form larger ones;sometimes with yellowing around the spot. *Alternaria* spp. cause brown to black and irregular leaf spots with concentric rings. Pea Nut Bud Necrosis Virus causes die-back symptoms on plant and ring spots on tomato -fruits.

Fig. 7.2: Cercospora Leaf Spot in Brinjal, Chilli and Spinach

Cercospora spp. initial spots are as chlorotic lesions, which are angular to irregular in shape.Later spots turn greyish-brown with profuse sporulation at the centre as black dots.

vi)   Blight: Most of the *Cercospora* spp. and *Alternaria* spp. infect either centre or margin of the leaf- lamina but in case of *Xanthomonas campestris, Phomopsisvexens, Phytophthora* spp. and *Stemphylium* spp. infection may generally start from the leaf margin. The blight symptoms are generally severe at the later stage of leaf spot diseases. Several pathogens depending upon the host-plant cause blight.

vii)   Powdery Mildew: It occurs on most of the plant species during dry season. Initially disease symptom appears as tiny yellow-greyish-white spots on the upper surface, which turn into necrotic

Fig. 7.3: Phytophthora Leaf Bligt (Late Blight) in Tomato

spots. Powdery growth occurs on both the leaf surfaces, which later turn into necrotic spots. Curling and premature defoliation, under- sized fruiting may also occur, plant growth is noticeably hampered. Powdery mildew infection is common in okra, cucurbits, brassicas, chilli, tomato, garden- pea, coriander, turnip etc.

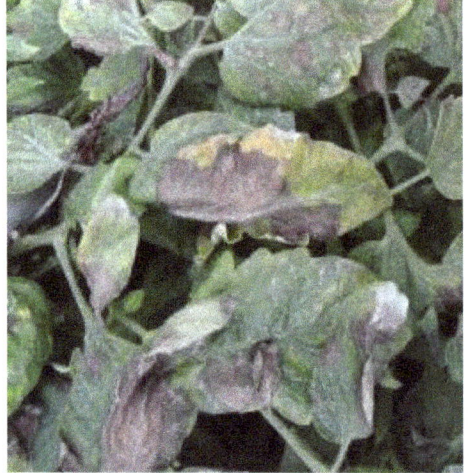

Fig. 7.4: Powdery Mildew in Bottle Gourd and Okra

viii) Downy mildew: It is common especially in cucurbits and brassicas. Symptoms occur as angular yellow spots on the upper leaf surface. Later spots turn into necrotic spots. Corresponding undersides of these spots are covered with greyish, mouldy growth. Infected leaves wither and die. Curling and defoliation may also occur. Plant growth is noticeably hampered.

Fig. 7.5: Downey Mildew in Ridge Gourd

ix) Rust: Initially causes yellow spots on both the leaf surfaces; underside gets covered with raised rust pustules. Disease is important in garden- pea. In case of white rust of brassicas, scattered pustules on lower leaf surface occur with white powdery growth mass. Floral organs show swelling and malformation. Inflorescence axis and stalks are thickened (12-15 times), petals and stamens fall off early and floral organs become swollen and fleshy.

Fig 7.6: Rust Disease on Garlic

x) Fruit-Spot and Fruit-Rot: It causes spots of varying colours, depending on the crop. The spot size also varies with the pathogen and host. In case of fungal diseases, fungal growth can be seen on the diseased spots.

Fig. 7.7: Alternaria Fruit Rot in Tomato and Anthracnose in Capsicum

7.2.2 Viral Diseases

Produce several symptoms like leaf- spot, upward and downward curling of leaves, vein clearing, stunting, die -back, chlorosis, mottling, deformation, and sometimes yellow rings, depending upon the virus and host- plants. Sometimes two or more viral infections can be on the same plant and may cause mixed symptoms. In this case, no visible growth of the organism can be seen on the infected parts, media and even under simple microscope. Sometime viral disease symptoms may be similar to insect injury.

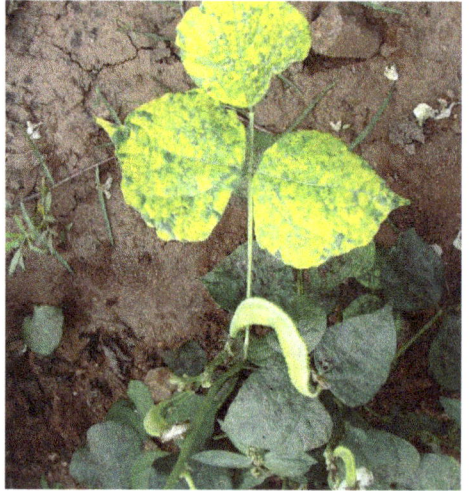

Fig. 7.8: Viral Mosaic Disease in Pumpkin and Lablab Bean

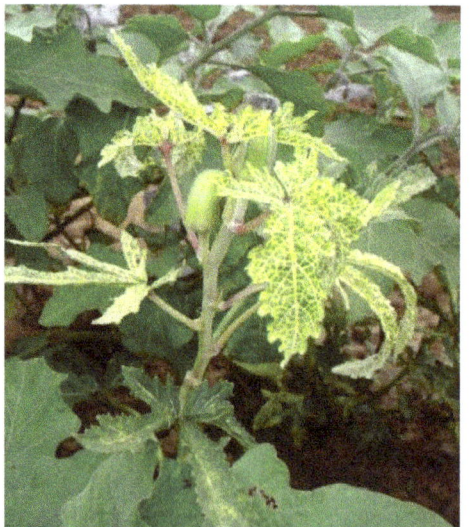

Fig. 7.9: Leaf Curl Viral Disease in Chilli and Yellow Vein Mosaic in Okra

7.3 Common Insect-pests in Home Gardens

a) Grasshoppers

In grasshoppers (pink, brown and green), crickets, praying mantises (green and brown) and cockroaches, forewings of the adults are usually long and narrow and somewhat thickened. Hind wings are membranous, broad and folded beneath the forewings at rest. Mouthparts are chewing type; antennae are often long and slender. Mole cricket and two-spotted cricket feed on plant roots and may be a problem in some cases. Praying mantises are general predators and feed on other insects.

b) Thrips (Thysanoptera)

They are small (very difficult to see by naked eye), slender insects with mouthparts modified into a short beak to suck plant sap. Their wings are slender with fringed margins. Thrips are important plant pests; their feeding often causes a stripling of leaf tissue accompanied by scarring, bronzing, or silvering. Some are major vectors of plant viruses. Melon thrips are pale yellow; are found on flowers and young foliage. They damage a range of plants, including cucumber, watermelon, tomato, eggplant and beans. Thrips are important vectors of tomato spotted wilt virus, affecting tomato, pepper, lettuce and flowering plants. In red-banded thrips, adults are black, while larvae are yellow with a red band on the abdomen; their feeding often scars fruits.

c) Bugs (Hemiptera or Heteroptera)

They are "True bugs," and the basal portion of their front wings is somewhat thickened and leathery; tip portion is membranous. The hind wings are membranous, and wings are held flat over abdomen with the tips of the front wing overlapping. They have piercing-sucking mouthparts formed into a slender beak. Some are feeding on plants, while others are predatory.

Stinkbugs are pests on beans, tomato and cabbage. Nymphal stages are dark coloured with whitish markings; adults are mostly light green and shield-shaped.

Fig. 7.10: Aphids Attack on Plants

Black stinkbugs are small, round and shiny black with pale stripes; they are an occasional pest on beans and some other legumes.Lace bugs cause stipling of leaves similar to other sucking insects; they commonly infest azaleas and rhododendron.

d) Aphids (Homoptera)

They are small, rounded or pear-shaped, soft bodied, mostly with a pair of tube-like cornicles on the posterior of the abdomen. Some are covered with a white powder. Aphids suck plant sap from leaves, stems and roots, often causing stunting, wilting and deformed leaves. Aphids excrete honeydew, a liquid high in sugar, which attracts ants and results in the growth of sooty mold fungus on infested plant parts, which interferes with plant photosynthesis. The group is very important as vector; transmitting plant viruses.

e) Whiteflies (Hemiptera)

They are tiny; adults resemble white moths; immature stages look- like scale insects. Adults' wings are covered with a white, waxy powder, making them impervious to wetness. Some are vectors of plant viruses and others cause various plant disorders.

Fig. 7.11: Whitefly Attack on Plants

f) Scales

They have adult females that are wingless, often legless, and sedentary. They are grouped under soft scales and armoured scales. Scales are soft- bodied, slow moving or sedentary, forming colonies with wingless forms. Adult's wings are held roof-like over the body; the antennae are often short and bristle-like (as with leafhoppers)

g) Mealybug (Homoptera)

Its females are oval and segmented with well-developed legs. Body is covered with a mealy or waxy substance. Mealybugs can be found on almost any part of the host-plant including leaves, stems, roots, and fruits. They spread by wind or are carried by ants, which feed on honeydew and protect insects from natural enemies.

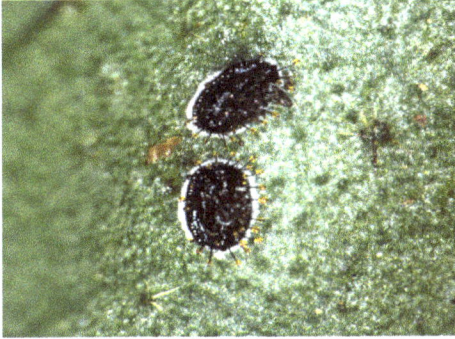

Fig 7.12: Scale Insect Attack on Plants   Fig. 7.13: Scale Insect Attack on Plants

h) Leafhoppers

They are elongated, slender insects with bristle-like antennae; wings of adults are held roof- like over the body, and they often hop when disturbed. They have one or two rows of spines on the hind legs. Some are vectors of plant viruses; others cause phototoxic reaction called hopper- burn.

i) Plant Hoppers

They are similar to leafhoppers but have a flattened spur on hind tibia and lack rows of spines on hind legs. Many have reduced or shortened wings

j) Termites (Isoptera)

Termite feeds on cellulose; found in plant material. Although normally found in wood, termites can feed on live plant tissues, including roots and fruits.

k) Insects with Compltamorphosis Coleoptera (beetles and weevils)

They are the largest insect order, including pests and beneficial insects. Its adults have a hardened, sometimes horny outer skeleton, usually with two pairs of wings, outer pair thickened, leathery, or hard and brittle, usually meeting in a straight- line down the middle, and the inner pair is membranous (mostly). Adults usually have a noticeable pair of antennae, variously shaped. Both adults and larvae have chewing mouthparts. **Beetle larvae**, also known as grubs, have a head capsule, three pairs of legs on the thorax, and no legs on the abdomen. **Weevil larvae** lack legs on the thorax. Foliage feeders, feed at night, and heavy infestations cause lace-like appearance of leaves. **Rose beetles** are common and damage many different plants including beans, eggplant, corn, cucumber, ginger, and ornamentals. **Tobacco flea- beetles** are tiny brown beetles and their feeding damage causes shot-hole appearance of leaves in eggplant and tobacco. **Stem borers** include long-horned beetles; its adults have long antennae and larvae bore into stems and wood. **Pinhole borers** leave pin-holes in branches and wood. **Fruit weevils** include pepper weevils; adults and grubs of which infest peppers and cause internal damage and premature drop. **Beneficial beetles** include ladybird beetles, also called ladybugs, which feed on homopteran insects such as aphids, scales, mealybugs, whiteflies, and psyllids and scavenger beetles, which help remove carcasses from the environment.**Lepidoptera (butterflies and moths)** have a caterpillar (larval) stage that causes most damage by chewing and boring, while adult, fruit piercing moth may be a pest on some ripe fruits. Most adult lepidoptera have long, siphoning, tube-like mouthparts to feed on plant nectar. Larval (caterpillar) stages have chewing mouthparts; most have three pairs of thoracic legs and five or less pairs of abdominal prolegs. Most larvae feed on leaves by leaf mining or bore into stem and fruit. **Diamondback moth** adult males have a diamond pattern on the wings when folded over the back. Diamondback moth is a pest of cabbage, and leek moth attacks onions. **Hawk moth caterpillars** are called hornworms for their distinctive, hornlike protrusion at the rear of the abdomen. Other pests include sweet-potato hornworm, imported cabbage worm and cabbage webworm.

l) Diptera

This includes flies, fruit flies, leaf miners and midges. Adults have only one pair of wings and have sucking mouthparts, which may modify. Their larvae are called maggots, are legless, and many lack even a well-defined head capsule, with only hook-like mouthparts. The order includes fruit flies, mosquitoes, house flies, horse flies and blow flies.

m) Leaf miners

Small adults lay eggs on plant tissues and larvae bore into tissues and create tunnels or mines. Heavy infestations can cause reduced photosynthesis and leaf drop, interrupt uptake of water and nutrients, and cause wilting. The group includes bean fly, serpentine leafminer, and vegetable leaf miner. Beneficial flies include parasitic flies like tachinid flies and predators like syrphid fly larvae and aphid flies; others are important as scavengers.

n) Mite pests

They are more closely related to spiders than insects, but some are important plant pests. Most mites are very small and difficult to be seen without magnification. Spidermites feeding damage includes stippling of leaves. Broad mites are found on many plants including tomato, beans, eggplant, kangkong and pepper, and they feed on young, growing leaves, causing distortion and bronzing.

o) Hymenoptera

Among ants, bees and wasps is an important group of plant feeders and includes beneficial pollinators, parasitoids and predators; used in biological control of insect-pests.

Insect can be beneficial: Not all insects are pests; they pollinate flowers producing fruits, seeds; produce silk, beeswax, shellac, honey and dyes; are used in biological control as predators and parasites to destroy pest insects and weeds; are food sources for some people, fish, birds, and animals, scavenge to remove carcasses, dead plant material, and dung ,help improve soil by burrowing and providing organic matter; are important in scientific research and genetics; can be pleasing and entertaining— some butterflies and beetles are colourful and are collected as a hobby, have had some value in medicine (such as maggots cleaned out wounds, honeybee stings for arthritis).

7.4 Integrated Pest Management

Integrated pest management (IPM) is a sustainable approach for managing pests. It combines biological, cultural, physical, and chemical tools in a way, which minimize economic, health, and environmental risks. It focuses on the use of non-chemical controls and selective use of chemicals when necessary. Weeds can be controlled by hand pulling or hoeing, and bugs can be removed by picking them off from vegetable and garden plants. Cleaning up dead leaves and debris removes potential homes to pests. Keeping disease- free seeds and soil, including beneficial insects and microorganisms into your garden are used to control pests.

The objective of IPM is to eliminate or reduce potentially harmful pesticide use by using a combination of control methods that would reduce pests to an acceptable level. The control methods should be socially acceptable, environmentally safe and economically practical. Many commercial agricultural systems use IPM methods to manage common insect -pest problems, and home gardeners can also use similar methods to control pest problems in the gardens.

7.4.1 IPM Components and Practices

IPM is an approach that emphasizes prevention of pest problems and use of least-toxic controls. The elimination or reduction in pesticide use can be achieved through thoughtful application of IPM strategies. IPM programmes follow a decision-making approach for managing pests, starting with making sure pests are correctly identified. In an IPM programme, treatments are only applied according to the need, in contrast to regularly scheduled applications or "calendar" sprays. One or more types of treatments may be combined in an IPM programme to provide desired level of control and to prevent future pest problems.

The IPM includes the following elements and factors.

a) Planning

Planning and managing ecosystems to prevent organisms from becoming pests;

b) Identification

Identifying pest problems and potential pest problems; proper identification of the pest or early detection help applying proper treatment

c) Monitoring

Monitoring populations of pests and beneficial organisms, damage caused by pests and environmental conditions; to know which stage of the pest causes damage and which pests are most susceptible to management with various possible control methods with an understanding of the pest life- cycle and its relationship to susceptible host- plant.

d) Action Decisions

Using injury thresholds in making treatment decisions; help identifying pest population at which treatment needs to be applied.

e) Treatment

Knowledge of the type of control methods available, and or alternative control methods; suppressing pest populations to acceptable levels using strategies based on considerations of the following.

i)   biological, physical, cultural, mechanical, behavioural and chemical controls in appropriate combinations, and

ii)   environment and human health protection.

f) Evaluation

Evaluating effectiveness of the pest management treatments helps in identifying control method to be followed in future if the same problem again occurs.

The important factors for applying any treatment including IPM depends on the following most important understandings.

Preparation: What control strategies can be used before you plant? Gardeners need to be aware of the potential problems and give plants the best chance to grow in a healthy environment. An important aspect of a successful pest management programme involves planning ahead to avoid pest problems as much as possible.

Monitoring: It consists of regular inspections to find out the extent of pest problem and whether it is getting better or worse. A basic monitoring programme would be a regular (daily or weekly) visual inspection coupled with written notes. Therefore, even a small amount of time spent on monitoring a problem can pay off by showing whether or not there is still a need for treatment. Problem should be detected whether pest numbers are low and easier to control, size of pest population or stage and the extent of the damage. Take a decision keeping in mind that the prevailing conditions should contribute to find out whether treatments are working for reducing pest problem. For most home garden customers, the emphasis should be on making frequent, close, visual examinations.

More complex monitoring programmes, using sampling tools, such as insect traps, are available commercially, which not only monitor pest population but also help reducing it. This includes the following traps.

g) Sticky Traps

Some insects are attracted to bright yellow or other colours so they can be caught on coloured sheets of plastics or cardboards that have been coated with sticky glue. Yellow sticky traps attract adult whiteflies, flower thrips, fungus gnats, leafminers and cabbage loopers. Bright blue traps attract flower thrips. Such sticky traps are usually used as a monitoring tool as well as to control pests. By regularly checking sticky traps, a gardener can find out when the first adult insect appeared.

Fig 7.14: Yellow Sticky Trap (YST) (a. Locally Made and b. Commercially Available) for Sucking Pests Like Whitefly, Leaf Hopper and Leaf Minor

h) Pheromone/kairomone traps

Adult female- insects, which are ready for matting, emit species-specific chemical odours to attract males and are called pheromones. This method of utilizing pheromone is called pheromone lure. Pheromone trap (Fig.1) attracts male moth and kairomone attracts female and male moths of some insects and are available for monitoring moth species. The pheromone traps septa/vials are baited/ impregnated with a lure, which mimics odour given off by female moths. The traps are used to find out when the main flights of adult moths occur so that spraying can be done at the right time to have greatest effect. Some pheromone traps are proven to control pest but it would not be useful for most home garden, because the gardener would not know how to relate number of moths caught to the need for treatments. Unfortunately, these traps alone would attract adult males at the time when damage has already been done by larvae. Since it is females that lay eggs, traps cannot control moth population.

Action decisions: Deciding when to take action and applying treatment is based on regularly monitoring pest situation. The treatments are made only when and where monitoring shows they are needed and not on a pre-set schedule. Injury level is the acceptable amount of injury or damage that could be tolerated from a particular pest population. Treatments are not automatically required when low numbers of "pest" species are found. Moreover, pest management decision not only depends upon the current monitoring of pest and disease occurrence but also on the prior history of a certain disease or a pest problem in a particular area or location. It includes deciding what types of problems would come and what types of precautions or treatments (cultural, biological, and chemical) are needed.

7.4.2 Treatments

For IPM system to achieve eco-friendly and sustainable insect and pest management, the inputs of both production as well as protection technologies are to be applied appropriately. IPM strategies consist of site preparation, monitoring crop and pest population, problem analysis, and selection of appropriate control methods. Crop- production technologies include: sanitation/clean cultivation, weed -free crop, summer ploughing, fallowing, flooding, crop rotation, integrated nutrient management, resistant/tolerant varieties, cultural practices like mulching, drip irrigation etc. Crop- protection technologies include: removal of crop plant residues/ debris, use of biocontrol agents, traps, botanicals and need- based pesticides. Broadly the treatments fall under two following major categories.

7.4.3 Prevention

The best way to prevent insect and disease problems is to maintain vigorous growth. Stressed plants suffer greater damage and succumb more quickly to pest infestation than do vigorous plants. Long-term goal of managing any pest problem should be prevention. This includes as follows.

Planting pest resistant plants: Many native plants may be naturally hardy and pest-free plants; and cultivars should be chosen, which have been selected for resistance. Always grow resistant varieties whenever possible. In some locations, certain disease organisms or insect pests exist in such quantities that production of specific vegetables would be more of a botheration.

Seed treatment: The seed is the base for healthy crop production. Always use disease- free seeds and do not save seeds if diseases are prevalent. Use of a protective fungicide/insecticide can help protect seeds against soil- borne diseases, especially during early stages of seed germination. Check new plants (before buying) to make sure they are not diseased or infested with insects. There are different chemicals and biological control agents used as seed treatments. Chemically treated seed should not be handled with bare hands.

Barriers to keep pests out: There are a variety of products, which keep away pests from crops. These include barriers such as floating row covers, net barriers, soil amendments and fertilizers, to grow healthy, pest -resistant plants. Barriers keep pests away from plants. Although they time and effort is required to install, once they are installed they usually require no further effort.

Cleaning up to remove food for pests: Cleaning up to remove food for pests: Cleaning of infected parts of the plants, destroying early dropped fruits, which are often infested with insects and destroying weed damaged/infected plant parts from surrounding area can keep home garden free from insect-pests. Removing food and water sources and eliminating nesting and over-wintering sites can also be very effective in controlling pest problems.

Cultural practices ensuring healthy plants growth:This can make a big difference in how well plants grow, their susceptibility to diseases, and how quickly they recover from damage. This entails selection of plants adapted to the conditions where they grow and physical destruction of individual pests on the host- plants. Sometimes large numbers of insect- pests can be destroyed with minimum effort—for example, hosing aphids off plants, or cultivating soil between rows to expose soil-inhabiting grubs and pupae to rays of sunand predation by birds. Turning plant residues in the fall allows them ample time to decay. The use of diseased plant material in a compost pile should be avoided. Keep weeds and fence rows mowed.

Crop rotation: Planting same crops in the same spot year after year decreases productivity. This is because soil- borne diseases, soil insects, nematodes, and toxic chemical residues tend to collect and build- up in a given area. As these detrimental factors increase, crop yields decrease. Therefore, it is necessary to rotate location of vegetables each season. Each family of vegetables has certain unique effects on the soil, and most vegetable varieties within a given family are susceptible to same diseases and insects. Therefore, it is important to know which vegetables are to be included in each family. Soil-borne diseases (for example, root rots and wilts) can be very destructive and difficult to control when populations of the disease-causing organisms are allowed to increase in the soil. Avoid planting same crops in the same area of the garden year after year to minimize these build-ups. Changes to the environment to make it less conducive to pests: When conditions that favour pest populations are removed or changed it can go a long way to control pests. This includes site selection. If possible home garden should always be in South direction and the planting rows should always be in east- west direction so that proper light is there for plant growth. It also includes planting crops in right direction (shorter plants towards south direction) and crop plot selection, selecting crop rotations, using slow- release fertilizers and correct watering practices. Aerating plants to improve drainage deter growth of moss and some species of weeds. Composting kitchen and garden wastes in rodent-proof bins reduces rodent populations. Sanitation would help reduce hazards of disease-causing microbes; being carried over from the previous crop. Plant refuse may be ploughed under in fall or removed from the garden and thoroughly composted before it is returned to the garden. Plant parts known to be diseased should be removed from the garden and not even composted, since pathogenic microbes may survive during composting. Plowing under plant residues in the fall or soon after clearing the garden space hastens decay of organic matter. Removal of diseased plants, plant residue, and weeds in and around the area would help stopping incidence of some diseases.

Preventive methods alone are often sufficient to keep most pest populations at tolerable level.

7.5 Control

Control measures are only called for where preventive measures have not been sufficient or some sudden insect-pest problems above the threshold level have occurred. Control measures for home garden pests generally fall under four main types: physical, mechanical, biological and chemical controls. These control measures may be used singly or in combination for their greater effectiveness. The control measures selected should be least hazardous to human health, least toxic to other non-target organisms (for example, pets, fish, and beneficial insects), environment-friendly, most likely to provide a long-term solution, most cost-effective in the long run to the customer and most likely to be used correctly by the customer

7.5.1 Physical Controls

Include mulching, sticky insect traps and rodent traps as well as using hoeing or hand pulling to control weeds. The application of mulches has many advantages, such as improving plant growth, enhancing the appearance of the garden, and reducing time spent in maintenance. It helps conserve moisture, maintaining a uniform soil temperature, reduce the weeds, furnish food for microorganisms found in the soil, preventing surface soil erosion, and improving soil texture. In addition, it helps keep leaves, flowers, and fruits free from soil which is important with strawberry and tomato. Straw, grass clippings, sawdust, leaves, newspapers, and black plastic sheeting all make excellent mulches for the home vegetable garden. If black plastic is used, it is best to cover it with lighter colored organic mulch to reduce the temperature around the plants. Mulches keep reducing the population of soil borne insect-pests as well as reduce the further spread of diseases and pests.

Sticky traps for insects can be used both as a monitoring tool and as a physical control. Yellow sticky traps are effective against a wide range of insects. They are practical when the pests are accessible and the numbers are not overwhelming— for example, handpicking tomato hornworms, individually destroying invading slugs, or pulling and destroying diseased plants to prevent the disease organism from spreading to adjacent healthy plants. Methods like removing and burning or burying insect-infested or diseased plant residues after harvest can help reduce these problems and hinder the pathogen-pest to over winter in the garden.

7.5.2 Mechanical Controls

Mechanical controls are machines or equipment used to control pests. These include vacuum cleaners, ultra-violet light traps, cultivators for weeds, and string weeders. Ultrasound repellers used to keep rats and mice away from gardens.

7.5.3 Biological Control

Biological control is the use of living organisms that are the natural enemies of pests and diseases to control their populations. A controlling agent can he an organism, a predacious or parasitic insect, predacious spiders or mites, or insect-feeding animals (rodents, snakes, toads) soil fungi and bacteria. Many microorganism bio-formulations are commercially available in the market like- *Bacillus species, Pseudomonas fluorescens Trichoderma species, Beauveriabassiana, Metarrhiziumanisopliae, Verticilliumlecanii, Gliocladium species, Nomuraearileyi, Nuclear polyhedrosis viruses (NPV), Trichogramma sp., Neem and plant based biopesticides* which not only control disease but also reduce the insect population.

Many of these beneficial organisms naturally occur in most gardens; however, effective numbers develop too late to control the pest organisms before some damage occurs. Populations of predators and parasites can multiply only after a sufficient food supply has developed for them. You cannot have insect-free garden produce and encourage beneficial insects at the same time. Releasing predators can be beneficial in establishing them in new areas. Home gardeners can employ biological controls by protecting the many beneficial species that are native to BC and buying beneficial species (insects or mites/ bioproducts) from commercial suppliers. The most important way to protect beneficial insect species is to avoid using pesticides as much as possible, especially chemicals that affect a wide range of species or that are very long lasting. Even if pesticides are needed as part of a pest management program, then select products should be used with the lowest toxicity, such as insecticidal soap, and/or shortest residual effect, such as pyrethrins. Applications should be limited to only those plants or areas that are infested.

Native beneficial insects can be attracted to home gardens by growing plants that provide them with pollen and nectar. Many adult female beneficial insects must have nectar or pollen to give them energy to lay eggs. Once the females are attracted to the garden, they are likely to stay and lay eggs. When the eggs hatch, the larvae that emerge prey on the pests. Plants with small flowers, such as dill, parsley (a biennial), thyme, and other herbs, which provide food for tiny parasitic wasps are good for insect attraction. Daisies are good pollen sources for lady beetles and other predators. Alyssum, candytuft, marigolds, wild mustard and salvias are also good insect plants. There are certain plants which can repel specific insect of other crops form the home garden. Marigold can repel cucumber and squash beetle and nematode, nasturtium deter aphids, beetle, squash bug, garlic repel beetle, aphid, weevils, spidermites.

Beneficial insects likes Asian lady beetle (*Harmonia axyridis)*, praying mentis for aphids, lacewings, ground beetles, leady bird beetles, and parasitic wasps, *Chrysoperla cornea, Trathalaflavo-orbitalis* for eggplant fruit and shoot borer, BDM larval parasite *Cotesia plutellae* and an egg parasite *Trichogrammatoidea bactrae* for tropical lowlands and larval parasitoid, *Diadegmase miclausum* and pupal parasitoid, *Diadromu scollaris* for highland areas *are* good biocontrol agents.

Commercial beneficial species sold by commercial suppliers include micro-organisms, insects, mites, and nematodes. The main micro-organisms currently available to gardeners are *Bacillus thuringiensis, Pseudomonas* spp. *Trichoderma* spp. *Trichoderma virance, Ampelomyces quisqualis, Penicillium* spp. *Lecanicellium lecanii, Beauveria bassiana, Paecilomyces licacinus, Nomuraea rileyi , Metarrhizium anisopliae, Hirsutella thompsonii* etc. are legally registered as pesticides.

While using beneficial species gardener should be aware that: most only attack particular species of pests, therefore the problem must be correctly identified first , supplier instructions must be followed very carefully to ensure good results and these living organisms which are very perishable and must be handled carefully and released promptly

7.5.4 Chemical Control

Managing pests is an important part of keeping your home garden vegetables healthy and productive Chemical control involves using a chemical pesticide to destroy pest organism or prevent it from spreading to other plants. Chemical pesticides can be used to control pests, but one should know that pesticides can cause water pollution and health hazards. In addition, studies have shown that home-owners often apply more pesticides than needed, and often incorrectly.

Chemical pesticides can be either organic or inorganic in origin. All pesticides can kill beneficial organisms as well as pests; therefore, the following precautions should always be followed. Be certain that your garden problem is caused by a pest organism and identify which one.

Follow label directions for application and observe all precautions listed on the product label.

Using no chemical pesticides and removing weeds, insects, and other pests by hand is the safest for environment and owner's health. If properly used, however, pesticides pose only a minimal risk. To use chemical pesticides safely, follow these five steps.

- Identify the pest or disease problem and determine the best way to control it.
- If need a chemical pesticide, purchase or use a pesticide formulation approved and labeled for control of the specific organism. Read the label to make sure that it is the right one for the job.
- Do not apply pesticides just it is to be done. Apply pesticide only on the affected crop.
- Apply the pesticides according to the label directions
- Carefully spray pesticides using good equipment to avoid contacting/spraying other plants and beneficial organisms.
- Be prepared to deal with accidental poisoning or spills
- Properly store and dispose of leftover pesticides and pesticide containers.

7.6 Select the Right Pesticide for the Treatment

To use a pesticide, one must read the label before applying, storing or disposing of. Different companies sell similar products under different brand names, so double-check is needed for buying the right product. Some pesticides contain chemicals designed to kill more than one pest; these are called "broad-spectrum" pesticides. Other pesticides have been developed to kill one or a few specific pests; these are called "selective" pesticides. Selective pesticides are best, because broad spectrum pesticides may kill beneficial insects also.

The equipment you need to mix and apply the product safely and effectively would depend on the individual chemical and the toxicity. The label also contains information on first aid, application restrictions, and other hazards. Labels, however, do not provide all the information you may need. You can ask your pesticide retailer for additional information.

Because pesticides can be hazardous; safe and proper disposal of any leftover chemicals can be time-consuming. The best way to avoid this problem is to purchase only the amount one needs to address the problem. There are several approaches to limit negative impacts of pesticides, such as: choose "preferred pesticides" where possible - choose ready-to-use formulations, rather than concentrates;choose application methods that limit the amount of pesticide used; limit treatments only to the plants or sites where pests are a problem; use alternative pesticide sprays if needed; notify anyone who enters a treated area, to check exposure to pesticides.

Apply the pesticide following label directions: Always follow label directions when you use any pesticide.

Be prepared to deal with accidental poisoning and spills: The pesticide label provides information on the relative toxic of the product.

- Chemicals labeled "CAUTION" have relatively low toxicity. Chemicals labeled"WARNING" have moderate toxicity.
- Chemicals labeled "DANGER" have high toxicity.

Depending on the chemical, pesticides can cause poisoning if they are swallowed, inhaled, get in someone's eyes, or, for very toxic chemicals, get on to someone's skin. Young children, who may not know any better, are at a higher risk for accidental poisoning. The label on the pesticide container contains basic information on first aid if someone is improperly exposed to the pesticide.

Properly store and dispose of pesticide containers: Pesticides should be stored in their original containers in a dry, well-ventilated, and secure (locked) location as per specific storage guidelines mentioned on the label. Leftover pesticides that are no longer needed must be disposed of as hazardous waste. Empty containers should be rinsed with water (triple-rinsed) and the rinse water should be disposed of on a vegetated area far from streams or ditches. The containers should be punctured and placed in your household trash for disposal.

Be warning: Many insecticides kill bees. Avoid spraying insecticides on plants that are surrounded by blooming flowers or weeds. Mow lawns next to the garden area to remove clover blossoms before applying any material hazardous to bees. This is a simple and important step. In all cases, when plants in the infested area are in bloom, select the material least hazardous to bees. Avoid using dusts whenever possible;sprays are preferred for bee safety.

Evaluation: Evaluation step in the IPM process is important because it helps people decide what worked for them and what didn't, so they can identify ways to improve results. In particular, the gardener should keep records of pesticides used and treatment dates. They should be encouraged to keep notes on the pest situation before they treat so that it can be compared with observations made after treatment. These notes would help them in future to decide whether or not treatments are necessary and when to take preventive steps.

7.7 Pesticide Handling

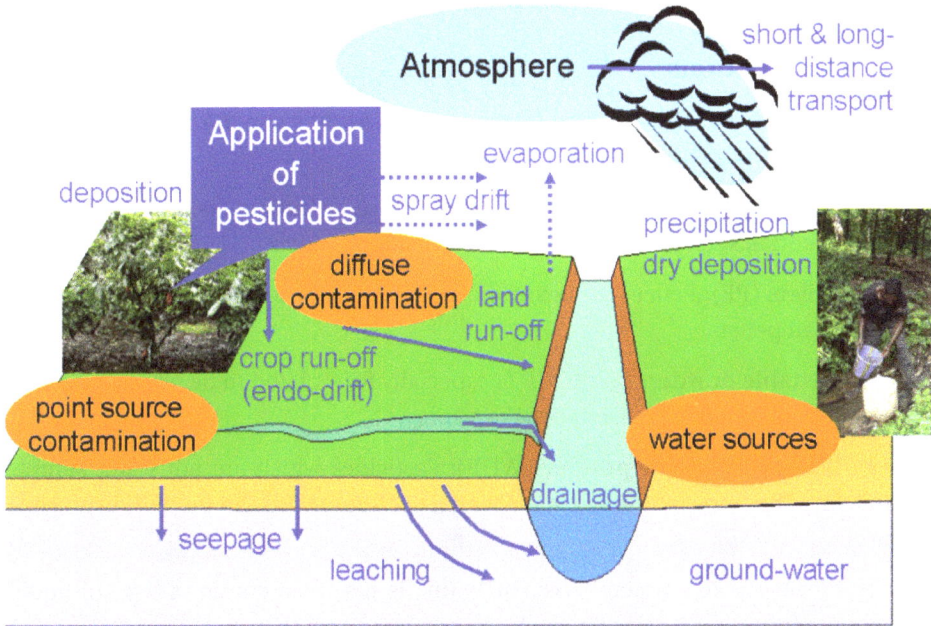

Fig 7.15: Pesticide Handling

7.8 Types of Pesticides

Pesticides are grouped according to the type of pests they are used for. Thus insecticides are for controlling insects, and herbicides are for controlling plants, fungicides control fungal diseases, and bactericides control bacterial diseases. Some pesticides are broad-spectrum (or non-selective) and others are selective pesticides.

Moreover, chemically pesticides are divided under several groups as their chemical constituents. How a pesticide works on the pests is called the mode of action. Some pesticides are stomach poisons, meaning pest is affected through feeding. Others act upon contact, meaning the pest dies if it comes in contact with the pesticide. How long the pesticide remains active against pests after it is applied is called its residual effect. Some pesticides remain active for days or weeks, while others, such as soap or oil sprays, act only on pests hit by the spray and have no residual activity after the spray dries out. Pesticides with residual effects are also persistent, meaning they remain active in the environment for a period of time, even if rain or wind has carried them away from the target site.

A few pesticides have systemic activity. This means that the pesticide is absorbed into the plant and moves through the plant tissues. A systemic insecticide can move throughout a plant to kill sucking insects on the leaves, while asystemic fungicide applied to the leaves/seeds/roots can move down/up to the roots/leaves to protect plants from foliar and or soil- borne diseases.

a) Formulations

i)   Dusts (D or DU): Dust products are used as they come from the container and are not mixed with water.

ii)   Granules (G or GR): Granules are incorporated into tiny beads made of clay or other material, such as mixed fertilizer-herbicide products.

iii) Pellets (PE): Pellets are incorporated into larger pieces than granules, such as rodent bait pellets.

iv) Wettable powders (WP): WP is especially formulated to be easy to mix into water and is intended to be used to make a spray.

v)   Water dispersible powder (WDP): Pesticides which are not water soluble but mix into the water and make a dispersible solution with the help of dispersible agent(s)

vi) Emulsifiable concentrates (EC): This is the most common type of liquid concentrate formulated so that it can be mixed with water to make a spray.

vii) Aerosols (A): These are ready-to-use sprays in small, pressurized containers, such as wasp and hornet sprays.

b) Storage Requirements

The amount of pesticides to be stored should be kept to minimum. Ideally, the customer should buy enough for one application or, at the most, enough for only one year. Purchasing excess pesticide requires long-term storage, which increases the risk to people, pets and the environment. Also, some pesticides degrade over time and lose their effectiveness. Pesticide should be located where children would not be able to reach.

Cool and dryThe storage area should be cool and dry, and protected from sunlight, extreme heat and freezing temperatures. Inspect the containers and store regularly:Stored pesticides should be checked for damaged or leaking containers. They may poison people or pets as they walk through the spilled pesticide or otherwise come in contact with it. Leaking pesticides also could cause environmental damage.

If a container is found to be leaking, it should be placed in a larger container, and the new container must be clearly labelled with the pesticide name and active ingredients.

Keep a spill kit: A basic spill kit should be located near the storage or display site and it should contain absorbent material such as kitty litter, a plastic bucket with lid, a broom and dustpan, and decontamination materials.

Transporting pesticides:Pesticides must be transported with care to avoid the possibility of spills. Always keep pesticide dry because if a package gets wet, the pesticide may become caked or hardened. Protective clothing and equipment: Pesticides can enter the body through skin, eyes, nose or mouth. The major route of entry into the body, however, is through skin. Prevent exposure to pesticides by wearing suitable protective clothing and other equipment as required. The minimum protective clothing worn while applying pesticides should consist of long-sleeved shirt, long pants, shoes and socks and unlined chemical resistant gloves. Additional protection requires rubber boots, coveralls (disposable type), rain suit (spraying trees), waterproof hat, goggles and respirator.

Mixing pesticides: To avoid need to mix pesticides, customers should purchase ready-to-use pesticides when possible. Ready-to-use products come in a variety of formulations, such as bait pellets, liquids, granules, or dusts, and are generally less toxic than pesticides that require mixing. It is essential to follow recommended mixing procedures because: pesticide concentrates are generally more toxic than diluted sprays, and the correct dilution is required to ensure effectiveness; incorrect dilutions can burn plants or cause other damages.

Pesticide Mixing area should be well ventilated, well-lit and a safe area in which to work. No other people (especially children), or pets, should be allowed near the mixing area. Mix only the recommended amount of pesticide as described on the label. Use protective cloths, scissors or a sharp knife to cut open paper bags carefully — do not rip them open, hold all pesticides and containers below eye level while handling to prevent pesticide from accidentally being splashed into the eyes and be careful not to spill or splash liquids, or to create dust, when mixing rinse empty liquid pesticide containers with a small amount of water and put the rinse water into the spray tank.

Applying pesticides: The user must not eat, drink, or smoke when applying pesticides because pesticides on the skin or clothing may be consumed or inhaled accidentally. If a pesticide is spilled on the user, the user should immediately, remove contaminated clothing, wash, and put on clean protective clothing before cleaning up the spill. Do not apply a pesticide just before, during or after a rainfall and in the afternoon. During pesticide application, spraying outdoors should be done when there is little or no wind. If little wind is there, the wind direction should be kept always from back to front. This is the movement of very small spray or dust particles and must be prevented or reduced as much as possible. Use a coarse spray of large water droplets rather than a fine spray to minimize drift. Always slight Pesticide should be applied close to the target and when there is no wind, air temperatures are below 30°C and mostly use granular instead of dust formulations

Avoid killing beneficial insects and other non-target living organisms by limiting the pesticide application to only those plants or areas that require treatment and when fruit trees or other plants are not in bloom.

c) After application

Clean-up hands and face with soap and water. The user should take shower if there is any body contact with the pesticide. Contaminated clothing should be removed and washed separately. People should not be allowed to enter the treatment area until the pesticide is dry. Wait at least 6 hours before re-entry — the longer the better. Pregnant women, new-born babies and people with breathing problems should not enter the treatment area for at least 12-24 hours. If it is necessary to re-enter the area during this time, proper safety protection for hands and feet should be worn (for example, rubber boots and gloves).

d) Observe 'Days to Harvest'

After a pesticide is applied to fruits, vegetables or other edible plants, the crop should not be harvested until after the required waiting period, usually stated on labels as "days to harvest." If harvested too soon after applying pesticides, food crops can contain unacceptably high levels of pesticide residues. The minimum number of days that should pass between the time a pesticide has been applied and when the crop is harvested depends on the pesticide and the type of the plant.

e) Protecting Environment

Always check labels for warnings about effects on the wildlife. As a rule, pesticides, which are toxic to humans are also toxic to wildlife such as birds, raccoons and squirrels. Pesticides with the poison symbol (skull-and-crossbones) on the label are of particular concern. Pay attention to all label warnings. Since naturally occurring beneficial organisms provide most of the control of pests in the environment, it is important to protect them in order to prevent pest problems. Some insecticides will also damage plants if mixed or used incorrectly. Some crops are sensitive to some selective pesticides which caused damage to the crops.

7.9 Preventing Environmental Damage

The best way to avoid damaging the environment with pesticides is to use non-chemical controls wherever possible. If pesticides are necessary, the first choice should be the lowest toxicity or least persistent product. Other ways users can help to reduce the risk of environmental damage include: spot sprays, to target plants only, should be used instead of broadcast sprays wherever possible. Pesticide sprays and runoff must not be used in a way that contaminates groundwater, ponds, steams, or marine areas. Unwanted pesticides and spray mixtures must not be poured into household or municipal drains. Pets and wildlife should be kept out of treated areas. Broadcast sprays and soil drenches of toxic insecticides should not be used on plants or areas where birds or other animals are likely to feed after treatment. Granular formulations of pesticides are particularly hazardous to birds. Rodenticides and slug baits must not be placed where these are accessible to pets or wild animals.

To protect beneficial insects, use insecticides that are the most selective, meaning that their effect is limited to certain group of insects. For example, products containing *Bacillus thuringiensis* only affect caterpillars, but not other insects. Use short-lived insecticides such as those containing soaps and pyrethrins.

Herbicide sprays and vapours must not be allowed to drift onto desirable plants. For example, do not use non-selective herbicides such as those containing glyphosate on lawn weeds, as they will also kill the turf grass. Soil sterilant herbicides should not be applied near trees, shrubs or garden areas.

7.10 Pesticide Sensitive Areas

Pesticides can be particularly damaging if applied too close to sensitive areas (buffer zones), such as desirable plants, water ditches, ponds, streams, or lakes. To avoid contaminating such areas, applicators should leave a "buffer zone" between the treatment area and the sensitive areas. A buffer zone is an area of land between the sensitive area and the treatment area where pesticides are not applied. This zone must be wide enough to ensure that any overspray, pesticide drift, or runoff could not cross this zone and reach the sensitive areas. The following are examples of buffer zone recommendations for most pesticides used by home and garden customers:

a) One meter around gardens

To protect ornamental beds and gardens from herbicides, or to protect vegetables, fruits, and other edible plants from insecticides being used on nearby ornamentals, leave greater distances if a breeze is blowing toward the sensitive area (better yet, wait until there is no wind). Granular herbicides and granular herbicides mixed with fertilizers should be applied no closer than one metre out from the drip line of trees. The drip line is the outer edge of the canopy of leaves on a tree, where water drops from the foliage. This is where the tree usually has many roots and where herbicides would be most damaging to the tree.

b) Two Times Tree Height

Soil sterilants should not be applied within a distance equal to two times the height of the tree; measured from the base of the tree.

This is because tree roots can extend long distances. For instance, if a tree is 10 metres high, then the soil sterilant should not be applied within 20 metres (2×10 meters) of the base of the trunk.

c) 10 Metres from Water

Pesticides should not be applied near ponds, lakes, streams or other bodies of water. A buffer zone of at least 10 metres should be kept between the margin of the water body and the edge of the treatment area.

7.11 Pesticide Disposal

Empty containers that held concentrated pesticides should be rinsed with water and pour the rinse material into the spray tank when mixing the spray solution. The rinsed container can then be disposed of in the garbage. Paper bags or cardboard packaging containers must not be burned as the burning bags or cardboard will release pesticide fumes. No goods or other items should be stored in used containers and the containers should not be re-used for any purpose. Gardeners should be encouraged to use alternatives to pesticides or to buy only small amounts of pesticides that can be used up in a single season.

8

Utilization-Harvesting, Post-harvest Management

8.1 Harvesting

Having a home garden is beneficial; it's good because you can go outside, pick out what you have grown, and eat it right there. It is also advantageous because of fresh taste, flavour of the product, besides being healthier. However, much of these benefits can be lost if vegetables are not harvested at the proper stage of development. Different vegetables have different harvest times and that's where it gets tricky. Some important things to remember while harvesting vegetables are as follows.

- to follow instructions and time limit on when to harvest.
- not to nick, bruise, or break fruits and do not pick a rotting plant to store with other vegetables as it may cause a disease to spread.
- storing them correctly is necessary for food to stay good and not spoiled.

Harvest quality of the products from the home garden depends on the appearance, size, shape, colour, blemishes, texture, flavour, nutritive value and safety.

a) Soil Preparation for Best Quality Harvest

Deep well- drained, fertile soil with plenty of organic matter provides optimum nitrogen for development of vegetables colour, flavour, texture and nutritional quality, **Excess N results in** hollowing and weight loss of sweet- potato; hollow stems in broccoli; branched carrots.**High levels of P** increases sugar concentration of vegetables while decreasing acidity. **High levels K** gives better quality vegetables; increases vitamin C and improves colour.

b) Facts for Harvesting Vegetables

Most vegetables are harvested just before full maturity; for maximum flavour and the most pleasant texture.

c) Timing

Harvest early in the day, dry and cool weather is the best and it prevents wounds. Cool vegetables quickly and thoroughly. Quality is reduced by improper temperature, drying, mechanical injury and diseases. Premature harvest reduces amount of flavour compounds. Late harvest may result in a fibrous, less tender, bland or bitter crop.

Table 8.1: Days to First Harvest for Some Selected Vegetables

Vegetable	Days to First Harvest
Amaranth	30-40
Asparagus	2 years
Basella	40-45
Beans, bush	55–75
Beans, pole	65–95
Beets	65–80
Broccoli	65–100
Chenopodium	35-45
Cabbage (early)	60–90
Cabbage (late)	110–130
Carrots	60-90
Cauliflower	50–75
Corn	70–140
Cowpea	55-70
Cucumbers	60-75
Eggplant	80-100
Kangkong	30-35
Lettuce (leaf)	55-60

Contd...

Vegetable	Days to First Harvest
Muskmelon	90–115
Mustard (leaf)	35-40
Onions (green)	50–70
Onions (dry)	90–110
Peas	60–120
Peppers	90–110
Pumpkin	190–195
Radishes	25–40
Spinach	50–60
Squash (summer)	60–70
Squash (winter)	75-90
Tomatoes (staked)	65–110
Tomatoes (sprawl)	65–110
Turnips	60–70
Yard long bean	50-60

Harvesting vegetables (Fig. 8.1) may seem like a simple task. "just pick them when they look right". However, the goal is to harvest vegetables when they are at their best - the most tender and sweetest. A number of vegetables if not picked at the optimum time become stringy, woody, or tasteless, which zeroes the hard work put into planting and taking care of them.

Fig. 8.1

d) Ensuring Vegetables Harvested at their Optimum

Asparagus: Begin harvesting when spears are 15-20 cm tall and as thick as small finger. Snap them off at the ground level and new spears will continue to grow. Stop harvesting about 4-6 weeks after the initial harvest to allow plants to produce foliage and food. If the plants were started from root crowns, one may be tempted to harvest in second year, although it's advisable to wait until the third year. Plants that are started from seeds, require at least 3 years before the harvest of spears. The spears may be harvested over a 2 week period during third year. In the fourth and subsequent years, harvest may be extended to four or more weeks. Cut a spear about 2.5 cm below the ground when the length of the spear is 17.5 cm. This would give an edible shoot of 20 cm. Crowns may be easily damaged by knife, so some people prefer to break off spears at the soil level. Eat asparagus immediately, or store them by immersing cut ends in water and refrigerate. Once well established, this perennial vegetable can be harvested for a period of 6-8 weeks each spring.

i)   Beans (Snap): Pick before one can see bulging seeds. Check daily. It doesn't take long for beans to go from tender to tough. Green beans indicate maturity by smoothness and greenness of pods. Snap beans are ready to pick if they snap easily when bent in half. Once pods begin to turn yellow and beans start to give corrugated look, the snap bean would become tough. Yellow beans varieties should also be picked before they appear corrugated. Days to maturity for beans are: bush beans, 50-60 days; pole beans, 60-90 days; limas 80- 100 days.

ii)   Beets: One can harvest and eat green tops, which are thinned out of rows. Beets are really a matter of personal preference when it comes to right size for harvesting. They are ready any time after beet shoulders protrude at the soil line. Beets may be harvested as soon as they reach 2.5 cm in diameter. They are excellent at 5 cm, but over 7.5 cm they become tough. The tender inner leaves can be used for pot greens. Beet greens make excellent borscht (beet soup).

iii)   Broccoli: one shouldn't expect home- grown broccoli to get the size of supermarket heads. Harvest broccoli when individual buds in the clusters are still tight. Once buds open, broccoli have a stronger flavour. The first crop would consist of the central head with 12.5 cm stem. A second crop can be harvested for several additional weeks, consisting of side shoots that have 7.5-10 cm of stem. Broccoli should be picked every 3-4 days so that crop does not get out of hand (Surplus broccoli can be frozen once blanched).

iv)   Brussel's Sprout: The sprouts would mature from bottom up. One can begin harvesting them when they are at least 2.5 cm in diameter. Harvest by twisting off or cutting sprouts from the stem. Sprouts should be bright green and firm. If the weather stays cool, one can expect 100 of these "tiny cabbages" per plant.

v)   Cabbage: These may be harvested as small as 10 cm in diameter and may be continued until 15-25 cm. The heads should be firm at harvest, but don't wait too long. A prolonged harvest may result in split heads. Maturity may take 2-3 months, depending on the year. One would feel cabbage head solid when gently squeezed.

vi)   Carrot: Carrots are hard to be judged. The tops of the carrot will show at the soil line, and can be gauged when diameter looks right according to the variety. If the diameter looks good, chances are the length is fine too. But would need to pull one of them to be certain. Carrots can be left in the ground once mature. Tiny sweet carrots can be harvested at 7.5-10- cm long. The remainder can be allowed to grow to about 2.5 cm in diameter. Once carrots reach 4 cm or larger, they will be woody. Maturity time from seed to harvest is 65-80 days, depending on the carrot variety and environment .

vii)   Cauliflower: Its head is made up of small segments resembling cottage cheese curds. For the best flavour, these segments should be tightly packed, white or ivory without brown spots. When heads start forming, tie leaves together at the top to form a teepee. This would keep the sun out and stop the head from yellowing. Its transplant to harvest time is approximately 70-80 days or up to 100 days if the cool weather persists.

viii)   Corn/Maize: Harvesting corn at the right time is vital for flavour. Maturity can be tested by peeling down husks. Pop a kernel 5- cm from the top end of the ear with fingernail. If the fluid is watery, it is still too early, and would need to wait a few more days; if the fluid is milky, the corn is at the right stage for eating; but if its consistency is of toothpaste type,then the corn has gone starchy and would be best used as creamed corn or in chowders. The milky kernel lasts only for a few days so harvesting should not be delayed. After picking cobs, cool them as quickly as possible and store in refrigerator. Once cobs are picked, they immediately change sugars into starches, especially in warm temperatures. Corn requires 65-90 days to mature, depending on the variety and environment.

ix)   Cucumbers: Cucumbers should be picked while still green and about 7.5-cm long for sweet pickles, 15- cm for dills and 20- cm for slicing. Picking 4-5 times a week would encourage continuous production. Do not leave mature fruit on the vine. Cucumbers pass their optimum stage once they turn yellow, form tough skin and have tough seeds. Days to maturity vary between 55and 70 days.

x)   Eggplant/Brinjal: Slightly immature fruits taste best; the fruits should be firm and shiny. Cut them rather than pull them from plant.

Fig. 8.2

xi)   Garlic: Its tops will fall over and begin to brown when the bulbs are ready. Dig, don't pull, and allow them to dry before storing. It is best to simply brush off the dirt, rather than washing.

xii)   Kale: Kale- leaves can be harvested throughout the season. They should be deep green with a firm, sturdy texture. Its flavour is best in cooler weather.

xiii)   Kohlrabil/Knolkhol: For the best texture, harvest bulbs when they reached about 5-7.5 cm in diameter. The bulbs become tougher as they grow and age. Pull or slice at the base.

xiv)   Leeks: Harvest leeks when they are about 2.5 cm in diameter.

xv)   Lettuce (Head): Harvest once head feels full and firm with a gentle squeeze. Hot weather would cause it to bolt or go to seed rather than filling out.

xvi)   Lettuce (Leaf): Harvest outer leaves once the plant reaches 10 cm in height. Allow younger, inner leaves to grow. Leaf lettuce can be harvested in this fashion for most of the summer.

xvii)   Muskmelon: There are many varieties of muskmelon, but a general rule of thumb is that colour should change to beige and fruit should 'slip' from the vine when lifted. One should also be able to notice a sweet smell when fruits are ripe.

xviii) Onions: Onions can be dug once e tops have ripened and fallen over. Allow onions to dry in the sun.

xix)   Okra: It can be harvested by picking; if by gentle pricking fruits are tender and still green. Fruits harvesting continues from ground to upside.

xx)   Parsnips: Parsnips taste best if left in the ground until after a frost or two. They can be left in the ground over the winter and harvested in spring. In cold areas, they should be mulched for winter.

xxi)   Peas: The pods should appear and feel full. Peas are sweeter if harvested before fully plumped. Peas really need to be tasted to determine if sweet enough.

Fig. 8.3

xxii) Potatoes: They can be harvested when tops start flowering. Carefully dig at the outer edges of the row. For full- size potatoes, wait until tops dry and turn brown. Start digging from the outside perimeter and move in cautiously to avoid slicing into potatoes.

xxiii) Pumpkins: Once the pumpkins have turned to the expected colour, and vines have started decaying, check on to make sure skin hardens enough that even poking with the fingernail doesn't lead to cracking. Do not pick pumpkin too soon, as it would stop turning orange once its cut, and don't leave them out if a hard frost is expected.

xxiv) Radishes: They mature quickly. Shoulders of the roots pop out of the soil line. If left too long, they would become tough and eventually go to seed production.

xxv) Rutabaga: The bulbs should be 7.5 cm in diameter, generally about 3 months after setting out. Rutabagas can be mulched, left in the ground and dug- up as needed. Cold weather improves their flavour.

xxvi) Swiss chard: As with leaf lettuce, cut outer leaves and allow the centre to continue growing. Pull off outer leaves as they mature, making sure to leave a central tuft of leaves to maintain the plant. Chard leaves are best harvested when they are 15-25- cm long. and Swiss chard needs 45-65 days to mature.

xxvii) Spinach: Spinach goes to seed quickly. Harvest by cutting 10-15 cm long at the soil line before a flower stalk begins to shoot up. Spinach needs 45-65 days to mature.

xxviii) Squash (summer): Pick it young and check often. The skins should be tender enough to poke fingernail through.

xxix) Squash (winter): Its colour is a good indicator of maturity. When the squash turns the colour, it is supposed to be cut from vine. Do not let winter squash to be exposed to frost.

xxx) Tomatoes: Harvest tomatoes when they are fully coloured and slightly soft to touch. Gently twist and pull them from the vine.

xxxi) Turnips: The turnip shoulders should be about 5-6 cm in diameter at the soil line, when ready. Harvest once they reach maturity. Overripe turnips become woody.

xxxii) Watermelon: White spot on the bottom of the melon should change to a deep yellow when ripe. Some people can hear a change in sound made when melon is thumped with a finger. It should make a hollow sound when ripe; this is a skill that must be developed.

8.2 Utilization of Home-Garden Vegetables and Their Nutritional Role

8.2.1 Importance of Vegetables

In the present scenario of prevailing malnutrition, vegetables are magic food packed with powerful vitamins and minerals to be safe from major diseases in life. Vegetables are the edible part of the plants which comes with multiple benefits. Vegetables can alleviate poverty, improve health by providing essential micronutrients, and enhance learning and working capacities of the children and adults through improved diets and health.

a) Simple Tips to Eliminate Pesticide Residues

Wash produce with large amount of cold or warm tap water, and scrub with a brush when appropriate; do not use soap. Throw away the outer leaves of leafy vegetables such as lettuce and cabbage.

b) Remove Pesticides with Vinegar

Clean the fruits and vegetables whether organic or not. It is important to eat a variety of healthy fruits and vegetables every day. What many people don't know is that much of the produce we eat is treated with chemicals including pesticides. Even small dose of pesticides can adversely affect health and diseases are on the rise. Advanced research continues to link various chemicals to chronic health conditions. Pesticides have been linked to learning disabilities and even cancer. Knowing this, it is important to work to reduce overall toxic load of chemicals White vinegar is an easy, economical option for cleaning fruits and vegetables. Exposure to pesticides and bacteria on produce can be reduced drastically with a thorough vinegar and water wash. Experts have found that white vinegar and water wash kill 98% of bacteria and remove pesticides.

The general consensus is to use 1 part white vinegar in 3 parts of water, mixed in a spray bottle. Spray the produce with vinegar mixture, and even giving an extra scrub brush before rinsing with water can help clean produce of pesticides.

c) Good Green Habits for Washing Produce

- Mix 3 parts water to 1 part white vinegar (3:1) in a spray bottle.
- Spray on fruits and veggies to get rid of pesticide residues.
- Rinse with water after spraying.

Or

- Fill a bowl with water and add 1/8 to 1/2 cup of vinegar, depending on the size of the bowl.
- Place fruits and veggies in the bowl.
- Soak for 15 to 20 minutes.

d) Healthy eating

Cleaning fruits and vegetables is essential for health. Cleaning produce with vinegar helps kill bacteria to ensure fruits and vegetables safe for consumption.

Smooth skinned produce: Keeping a blend of vinegar and water at a 1 to 3 ratio in a spray bottle makes cleaning smooth-skinned produce easier. Use spray bottle to mist fruits and vegetables; thoroughly coating its exterior with vinegar solution. Allow produce to be like that for 30 seconds before rubbing its surface and rinsing it under cold, running water. This removes all vinegar taste. Clean smooth-skinned fruits and vegetables by gently rubbing them with hands instead of an abrasive scrubber. This is necessary to protect the skin before the fruit or vegetable is completely clean, which could expose flesh to contaminants. Tomatoes, apples and grapes are examples of smooth-skinned produce.

Rough or firmed skinned: Broccoli, cauliflower, leafy greens, melons, potatoes, berries and other produce without a smooth or soft surface are slightly more difficult to clean. They require a soaking in 1 to 3 vinegar and water mixture. This ensures acidic blend killing all bacteria. For heads of cabbage or other greens, one needs to separate individual leaves for thorough cleaning. This can be a bit impractical, but sink in the home can be used as the container for water and vinegar mixture. After soaking, scrub vegetables with a brush and rinse them under running water.

Other precautions: To stay safe when cleaning fresh fruits and vegetables, wash your hands thoroughly before and after handling them. Also, thoroughly wash any surface they touched, including knives and cutting boards. Never cut or peel fruits and vegetables before washing them, as this can contaminate flesh. Always dry produce with a clean -cloth and cut away damaged areas before serving. When working with cabbage and lettuce, discard outer leaves and wash inner leaves.

Tips: Adding lemon juice to the mix can boost acidity, which may help kill more bacteria. Blending lemon juice with vinegar mixture makes it more effective by increasing acidity. This can help kill more bacteria, including *E. coli*. Washing berries with a vinegar solution offers additional benefits -- it prevents them from moulding within a few days of purchase. When shopping, choose unbruised and undamaged produce.

8.2.2 Utilization of Vegetables

a) Salads:

uncooked as side dishes, Cooked dishes, Boiled, sautéed, curried, fried, broiled or baked for consumption, Garnish for other dishes especially meat and fish, Canned, pickled or dehydrated, Frozen for future use **and** Some beans such as mungbean and soybean can be processed into bean cakes (tofu) or fermented to tempeh

Table 8.2: Vegetables Suitable for Different Types of Processing

Processing Type	Vegetables
Canned	Okra, asparagus, tomato, kangkong, sweet- corn, potato, sweet-potato, carrot, snap bean, bamboo- shoots
Pickled	Cucumber, cauliflower, bittergourd, onion, carrot, turnip, radish, sweet pepper, ginger, chilli, garlic
Dehydrated	Onion, garlic, carrot, bell pepper, potato, sweet- potato, mustard, ginger
Fermented	Cabbage, Chinese cabbage, radish, mustard
Other products	Potato (fries, crisps, flour), carrot (jam), mungbean and soybean (tofu and tempeh)

Table 8.3: Parts of Plant Used as Vegetables

Plant Part	Vegetable
Bulb	Onion, garlic
Root	Radish, carrot, sweet- potato, turnip
Tuber	Potato
Stem/rhizome	Bamboo shoot, asparagus, ginger, taro
Leaf	Cabbage, lettuce, mustard, spinach, kangkong, basella, coriander, parsley, lettuce, amaranth
Fruit	Cucumber, eggplant, tomato, pepper, beans, pea, gourds, okra, corn, melons
Flower	Cauliflower, broccoli

b) Vegetables with Medicinal Properties

i)   **Tomato:** For skin protection; improve skin ability to protect against harmful UV rays; reduce risk of prostate cancer; have curative effects in low-appetite; stomachache; controlling excessive fat and blood impurities; cure piles, jaundice, weakness and fever. Okra: Helps stabilize blood sugar by curbing the rate at which sugar is absorbed from intestinal tract; helps prevent and improve constipation; used for healing ulcers and to keep joints limber; treats lung inflammation, sore throat, and irritable bowel; a supreme vegetable for those feeling weak, exhausted, and suffering from depression; mucilage binds cholesterol and bile- acid carrying toxins are dumped into it by the filtering liver.

ii) Cruciferous vegetables (broccoli, cauliflower, Brussel's sprouts, mustard greens, cabbage): Potent protective properties for mucous membranes, especially of lungs and digestive tract, and are, therefore effective guardians against cancer, ulcers and infections in these vital organs; rich in antioxidant nutrients such as beta carotene, vitamin C and selenium; guard body against all sorts of toxins absorbed from polluted environments.

iii) Ginger: Digestive properties

iv) Garlic: Inhibits growth of tumors; increases activity of white blood cells and macrophages; raises antibody production; provides a rich dietary source of rare trace element selenium, which is a potent antioxidant; destroys fungus and yeast.

v) Bitter-gourd: Used as a folk medicine for diabetes; antidotal, antipyretic tonic, appetizing, stomachic, antibilous and laxative; juice of fresh leaves of bitter gourd is valuable in piles; highly beneficial in the treatment of blood disorders like blood boils, scabies, itching, psoriasis, ring-worms and other fungal diseases.

vi) Spinach: Causes formation of urine; activates intestine; eliminates excessive bile and phlegm in the body; purifies blood; cures stone and strengthens bones.

vii) Brinjal: Good source of potassium and other essential nutrients; used for treating hypertension, cancer and diabetes; treats enlarged spleen caused due to malaria; maintains blood cholesterol level; cures insomnia; treats congestion and phlegm.

c) Vegetables of cosmetic value Cucumber

Countless health benefits as well as cosmetic properties- an excellent source of vitamin C, folic acid and potassium; skin of cucumber is rich in fibre and contains variety of minerals, potassium, magnesium and silica; prevents and treats under eye dark circles when slices of cucumber are kept on the closed eyes regularly before sleep in night; improves skin complexion and glow as it is rich in silica; used in all the face pack preparations because of its cooling, cleansing and diuretic properties as a best natural cosmetic for skin; slices of cucumber placed on closed eyes reduce swelling and soothe the eyes; helps preventing splitting of nails of fingers and toes ;relaxes and elevates sunburn's pain, heat and inflammation; its juice is said to promote hair growth, especially when added to the juice of lettuce, spinach and carrot.

8.3 Storage Methods for Vegetables

Basically, storage is placing harvested vegetables in an environment where life processes, respiration and water loss are kept at low levels. Many vegetables grown in home gardens can be stored fresh, but they must be harvested at the proper maturity and kept at the correct temperature and humidity. In addition, proper ventilation and sanitation must be maintained during storage.

8.3.1 Storage Condition

a) Respiration During Storage

During respiration, sugars and other compounds are broken down within cells. This releases energy, carbon dioxide, water and heat. The energy is needed by the living cells of the stored product. The carbon dioxide should be removed by adequate ventilation.

Several factors regulate respiration. In general, higher the temperature(within normal range), faster is the respiration rate. The presence of soluble sugars in the cells also influences rate of respiration. At 70 °F, respiration rate of sweet- corn is 3.6 times as fast as it is at 41 °F. Thus, it needs to be cooled immediately after harvest. Respiration rate also directly varies with water content. At a given temperature, succulent plant parts, such as head lettuce, respire more rapidly than non-succulent products. Immature vegetables respire more rapidly than mature vegetables. And finally, respiration rate is influenced by oxygen level. During respiration, oxygen is absorbed and carbon dioxide is released. Consequently, an airtight area would decrease in oxygen and increase in carbon dioxide. As the result, respiration rate gradually decreases. However, if an area is completely airtight and oxygen levels fall too low for complete combustion of sugars, undesirable compounds are produced, which would lower edibility of vegetables. Therefore, respiration should be held at low level rather than be stopped completely. For this reason, vegetables and fruits are often wrapped in perforated plastic containers in supermarkets.

b) Water Loss During Storage

Water loss in fresh vegetables results in wilted and dull appearance and thus reduces their eye appeal and freshness. Containing water loss improves shelf- life, appearance and desirability. The loss in storage is prevented by storing product at as low a temperature and as high a relative humidity as possible. Most commonly vegetables are stored in a refrigerator.

Refrigerators usually maintain a temperature of about 40 ° F, but temperature may vary within the storage compartment. In single-door models with a frozen food storage unit, temperatures are generally lowest just beneath the storage unit. Cold air settles and forces warm air near the vegetable tray upward along the sides. The circulation air is usually of lower humidity and would dry out uncovered vegetables. However, humidity in the tray can be maintained at a higher level by using moist towels or by abundance of vegetables. While many vegetables can be stored well in the refrigerator for a week or longer, certain storage precautions need to be observed. For instance, many ripening fruits should not be stored together with vegetables. These fruits give off ethylene gas, which causes yellowing of green vegetables, russet spotting on lettuce, toughening of asparagus spears, sprouting of potatoes and bitterness in carrots. Some high ethylene producing fruits are pears, plums, apples, cantaloupes and peaches.

8.3.2 Refrigeration

Different fruits and vegetables should be stored in different way. Vegetables generally need one of four types of storage. These are: cold (32-39° F), moist storage; cool (40-50 ° F), moist storage; cold, dry storage; warm (50-60 ° F), dry storage. Typically, a refrigerator should be kept at around 34 ° F. Vegetables are best stored in crisper section of the refrigerator. This is the drawer or drawers located at the bottom of most refrigerators. Crispers usually have dedicated humidity controls.

- Apples, broccoli, carrots, lettuce and eggplant are best in high humidity and cool or cold temperatures.

- Garlic and onions are best in low humidity and cold temperatures

- Hot peppers, pumpkins, winter squash, sweet potatoes are best in warm, dry conditions

- Any vegetable that has been washed and cut should be refrigerated for safety. Store such vegetables in a plastic bag to preserve freshness and limitig their contact with air.

a) Freezing

Almost all fruits and vegetables can be stored in freezer. This usually does not reduce health benefits or vitamin contents. The freezer is a great way to store seasonal fruits or vegetables for use later in the year. It's best to freeze fruits or vegetables in airtight containers. Avoid freezing produce if it's not ripe; as it may not ripen correctly when out of the freezer. Leafy greens, such as lettuce and spinach, should not be frozen.

b) What Not to Refrigerate

Certain types of produce are best left out of refrigerator. Instead, they should be stored in a cool, dry place. These include tomatoes, bananas, potatoes, lemons and limes. Tomatoes, in particular, are known to lose flavour and nutrients if refrigerated. They also can develop an undesirable texture. Whole fruits generally doesn't need refrigeration. However, if refrigerated, it arrests ripening.

8.3.3 Vegetables According to Storage Requirements

Group I: Keep at 32 to 41° F and 85 to 95% relative humidity. Store in refrigerator crisper and maintain high humidity by keeping crisper more than half full. Wash and drain well vegetables before storage. This group includes: beet greens, Swiss chard, collards, green onions, kale, lettuce, mustard greens, spinach, and turnip greens. And store the following vegetables in a crisper separate from above vegetables in plastic bags or containers in the main compartment of the refrigerator: asparagus, beets, broccoli, cabbage, carrots, cauliflower, celery, lima beans, peas, radish, sweet-corn (if un-husked, keep close to freezer compartment) and turnip.

Group 2: Ideally, the following vegetables remain best at 45 to 55 ° F and 85 to 95 % relative humidity due to their sensitivity to chilling injury. Since this is not always possible in most homes, store in refrigerator no longer than about seven days. Use soon after removing from refrigerator. The group includes: bell peppers, hot peppers, cucumbers, ripe melons, snapbeans, and summer squash.

Group 3: Store vegetables of this group in a cool place (50 to 60 degrees F). Lower temperature may cause chilling injury. Pantries, basements or garages can provide a cool place during most of the year. Non-insulated garages may be too warm in summer and too cold in winter. This group includes: eggplant, okra, ripe tomatoes, potatoes (store in subdued light to prevent greening), sweet -potatoes

Group 4: Store the following vegetables at room temperature (65 to 70 ° F) and away from direct sunlight. They are: dry garlic, melons, dry onions (in open mesh container), tomatoes (mature green, partly ripe and ripe)

8.4 Vegetable Preparation/Cooking

Food preparation is an important step in meeting nutritional needs of the family. Food has to be pleasing in appearance and taste to be consumed. Foods like fruits, some vegetables and nuts can be eaten raw but most vegetables are cooked to bring about desirable tas.te and change The process of subjecting food to the action of heat is termed as cooking.

Objectives of Cooking

Cooking sterilizes food. Above 40° C growth of bacteria decreases rapidly. Hence food is safe for consumption. Cooking softens connective tissues of coarse fibres of cereals, pulses and vegetables so their digestive period is shortened and gastrointestinal tract is subjected less to irritation. Palatability and food quality improves by cooking – appearance, flavour, texture and taste of the food enhances while cooking.

8.5 Cooking Methods

8.5.1 Moist Heat Method

a) Boiling

Boiling is a method of cooking food by just immersing it in water at 100° C and maintaining water at that temperature till food becomes tender. Dhal, roots and tubers are cooked by boiling.

- **Merits:** Simple method - It does not require special skills and equipment. Uniform cooking can be achieved.
- **Demerits:** Continuous excessive boiling leads to damage in structure and texture of food. Loss of water- soluble nutrients such as B and C vitamins if water is discarded. Time consuming – Boiling takes more time to cook food and wastes fuel . Loss of colour – water soluble pigments may be lost.

b) Stewing

It refers to simmering of food in a pan with a tight- fitting lid using small quantities of liquid to cover only half the food. This is a slow method of cooking. The liquid is brought to boiling point and the heat is reduced to maintain simmering temperature (82ºC - 90º C).

The food above the liquid is cooked by steam generated within the pan. Legumes are usually stewed.

- Merits: Loss of nutrients is avoided as water used for cooking is not discarded and flavour is retained.
- Demerits: The process is time consuming and there is wastage of fuel.

c) Steaming

In this, food is cooked in steam generated from vigorously boiling water in a pan. The food to be steamed is placed in a container and is not in direct contact with the water or liquid. Steaming is one of the easiest and most nutritious way to cook vegetables. In steaming, vegetables are cooked gently over – not in – hot water. Vegetables are put in a perforated steaming basket which is then placed in a pot containing 1.5 to 3 cm of water. Water should be below the level of the steaming basket, so that nutrients aren't leached away into the water. Steaming is done over medium heat. Be careful not to let all the water evaporate away. Almost any vegetable can be steamed, including asparagus, spinach, summer squash, peas, broccoli, sweet-corn, string beans, Brussels sprouts, cauliflower, Swiss chard, kale, cabbage, beetroot, onions, potatoes, turnips, turnip greens, mustard greens and sweet- potatoes.

Dense vegetables should be cut into small pieces to cook more quickly.

- *Merits:* Less chance of burning and scorching. Texture of food is better as it becomes light and fluffy. Cooking time is less and fuel wastage is less. Steamed foods like idli and idiappam contain less fat and are easily digested and are good for children, aged and for therapeutic diet. Nutrient loss is minimized.
- *Demerits:* Steaming equipment is required. This method is limited to preparation of selected foods.

d) Pressure cooking

When steam under pressure is used, the method is known as pressure cooking and the equipment used is pressure cooker. In this method, temperature of boiling water can be raised above 100° C. Tubers are usually pressure cooked.

- *Merits:* Cooking time is less compared to other methods. Nutrient and flavour loss is minimized .Conserves fuel and time as different items can be cooked at the same time. Less chance for burning and scorching. Constant attention is not necessary.
- *Demerits:* Knowledge of the usage, care and maintenance of cooker is required to prevent accidents. Careful watch on the cooking time is required to avoid over cooking.

e) Poaching

This involves cooking in the minimum amount of liquid at 80° - 85° C; that is below the boiling point.

- *Merits:* No special equipment is needed. Quick method of cooking and therefore saves fuel. Poached food is easily digested since no fat is added.
- *Demerits:* Poached foods may not appeal to everybody as they are bland in taste. Food can be scorched if water evaporates due to careless monitoring. Water soluble nutrients may be leached into water.

f) Blanching

In meal preparation, it is often necessary only to peel off skin of fruits and vegetables without making them tender. This can be achieved by blanching. In this method, food is dipped in boiling water for 5 seconds to 2 minutes depending on the texture of the food. This helps remove skin or peel without softening food. Blanching can also be done by pouring enough boiling water on the food to immerse it for some time or subject food to boiling temperature for short period and then immediately immersing in cold water. The process causes skin to become loose and can be peeled off easily.

- *Merits:* Peel can be easily removed to improve digestibility.It destroys enzymes which bring about spoilage. Texture can be maintained while improving colour and flavour of food.
- *Demerits:* Loss of nutrients if cooking water is discarded.

8.5.2 Dry Heat Methods

a) Roasting

In this method, food is cooked in a heated metal or frying pan without covering, e.g., sweet- potato

- *Merits:* Quick method of cooking. It improves appearance, flavour and texture.
- *Demerits:* Food can be scorched due to carelessness. Roasting denatures proteins reducing their availability.

b) Grilling

Grilling or broiling refers to cooking of food by exposing it to direct heat. In this method, food is placed above or in between a red hot surface ;corn can be prepared by this method.

- *Merits:* Enhances flavour, appearance and taste of the product. It requires less time to cook. Minimum fat is used.
- *Demerits:* Constant attention is required to check charring.

c) Toasting

This is a method where food is kept between two heated elements to facilitate browning on both the sides.

- *Merits:* Easy and quick method. Flavour improves.
- *Demerits:* Special equipment is required. Careful monitoring is needed to avoid charring.

d) Baking

In this method, food gets cooked in an oven or in oven- like appliance by dry heat. The temperature range maintained in an oven is 120 – 260ºC. The food is usually kept uncovered in a container greased with a fat coated paper. Vegetables are prepared by this method.

- ***Merits:*** Baking lends a unique baked flavour to foods. Foods become light and fluffy. Certain foods can be prepared only by this method – bread, cakes. Uniform and bulk cooking can be achieved. Flavour and texture are improved. Variety of dishes can be made.

- ***Demerits:*** Special equipment like oven is required. Baking skills are necessary to obtain a product with ideal texture, flavour and colour. Careful monitoring needed to check scorching.

e) Sautéing

It is a method in which food is lightly tossed in a little oil just enough to cover base of the pan. The pan is covered with a lid and the flame or intensity of heat is reduced. The food is allowed to cook till tender in its own steam. The food is tossed occasionally, or turned with a spatula to enable all pieces to come in contact with the oil and get evenly cooked. The product obtained by this method is slightly moist and tender but without any liquid or gravy. Foods cooked by sautéing are generally vegetables which are used as side dishes in a menu. sautéing can be combined with other methods to produce variety in meals. Many vegetables are good sautéed in a little olive or canola oil (alone or in combination), including courgettes, yellow summer squash, spaghetti squash, leafy greens, onions, garlic, peppers, snow peas, string beans, broccoli and carrots. Dense vegetables such as broccoli or carrots may be sliced thin, steamed first to cut down on cooking time.

- ***Merits:*** Takes less time. Simple technique. Minimum oil is used.

- ***Demerits:*** Constant attention is needed as there is chance of scorching or burning.

f) Frying

In this method, food to be cooked is brought into contact with larger amount of hot fat. When food is totally immersed in hot oil, it is called deep -fat frying. Vegetable pakoras are examples of deep- fat fried food. In shallow fat frying, only a little fat is used and the food is turned so that both sides become brown, e.g., vegetable cutlets.

- ***Merits:*** Very quick method of cooking. The calorific value of food increases since fat is used as the cooking medium. Frying lends a delicious flavour and attractive appearance to food. Taste and texture also improves.

- ***Demerits:*** Careful monitoring is required as food gets easily charred when smoking temperature is not properly maintained. The food may become soggy due to too much of oil absorption. Fried foods are not easily digested. And repeated use of heated oil would have ill effect on the health.

8.6 Combination of Cooking Methods

a) Braising

It is a combined method of roasting and stewing in a pan with a tight- fitting lid

Flavouring and seasonings are added and food is allowed to cook gently.

b) Microwave Cooking

Microwaves are electromagnetic waves of radiant energy with wave lengths in the range of 250×106 to 7.5×109 Angstroms. The most commonly used type of microwave generator is an electronic device called a magnetron which generates radiant energy of high frequency. A simple microwave oven consists of a metal cabinet into which magnetron is inserted. The cabinet is equipped with a metal fan that distributes microwave throughout the cabinet. Food placed in the oven is heated by microwaves from all directions. Moist foods and liquid foods can be rapidly heated in such ovens. Food should be kept in containers made of plastic, glass or china ware which do not contain metallic substances. These containers are used as they transmit microwaves but do not absorb or reflect them.

- *Merits:* Quick method – 10 times faster than conventional method. So loss of nutrients can be minimized. Microwave cooking enhances flavour of food as it cooks quickly with little or no water.
- *Demerits:* Baked products do not get a brown surface. Microwave cooking cannot be used for simmering, stewing or deep- frying. Flavour of all ingredients do not blend well as cooking time is too short.

c) Solar Cooking

It is a very simple technique that makes use of sunlight or solar energy ; a non-conventional source of energy.

Solar cooker consists of a well-insulated box which is painted black on the inside and covered with one or more transparent covers. The purpose of these transparent covers is to trap heat inside the cooker. These covers allow radiations from the sun to come inside the box but do not allow heat from the hot black absorbing plate to come out of the box. Because of this, temperature up to 140°C can be attained, which is adequate for cooking.

- *Merits:* Simple technique – requires no special skill. Cost -effective as natural sunlight is the form of energy. Original flavour of food is retained. There is no danger of scorching or burning. Loss of nutrients is minimum as only little amounts of water is used in cooking.

- *Demerits:* Special equipment is needed. Slow cooking process. Cannot be used in the absence of sunlight – rainy season, late evening and night.

d) Baking and Roasting

Vegetables can also be baked along with a Sunday roast or in a casserole or baked pasta dishes. Sneak more vegetables in your diet by incorporating them into baked goods such as carrot cake, bread, pumpkin muffins, or sweet- potato pie.What used to be called roasting is now more commonly called grilling or barbequing.

e) Deep-fat Frying

The most popular deep-fried vegetable is potato – in the form of chips or crisps. Vegetable tempura. : Vegetables dipped in batter and deep-fried – is another great way to enjoy vegetables. Asparagus, eggplant, broccoli, carrots, cauliflower, onions, peppers, potatoes, squash, string beans and sweet -potatoes all make good tempura.

f) Stir-frying

It is a very healthy cooking method. It cooks quickly without water (which retains vitamins) and uses very little oil. Vegetables are cut into small pieces and added to a preheated wok containing a small amount of oil. The vegetables must be stirred constantly to ensure even cooking and avoid burning. The key to proper stir-frying is that keep everything chopped and ready to go. With gas stoves, a wok with either a round or flat bottom can be used. Round-bottomed woks are set in a ring that keeps them stable. On an electric stovetop, either a flat-bottomed wok or a skillet is used. Heat empty wok first; once the wok is hot add a little oil, rotate the pan to coat the surface. Good vegetables for stir-frying include broccoli, carrots, onions, peppers, garlic, peas, cabbage, string beans, cauliflower and leafy brassicas.

g) Boiling and simmering

Vegetables that are often boiled include potatoes, cabbage, root vegetables and onions. Many nutrients are boiled away into the water, so try to save liquid and add to soups and stews. Soups, stews, and dried bean dishes are simmered, not boiled, and since the liquid is consumed as part of the dish, nutrients are not lost.

8.7 Best Cooking Method

Pressure cooking is the best cooking method where nutrient loss is minimal and cooking time is also less. Foods cook fast and uniform if all are cut in uniform and medium shape.

Nutrient Loss

Water soluble nutrients are lost if more water is used for cooking and then discarded. Fat soluble nutrients are lost if more fat is used for cooking like in deep- fat frying. Enough water should be used to cook vegetables. If there is more water, that can be used to make soups and vegetable curries. Oxidisable nutrients like vitamin C is lost when vegetables are cut much in advance and kept exposed to air so vitamin c rich foods should be cut just before cooking or should be immersed in lime solution or water immediately after cutting till the time of cooking. Vegetables should always be washed before cutting to prevent nutrient loss.

8.8 Vegetable Cooking Tips

a) Stir-frying tips

Cut vegetables in same-size pieces so they'll all take the same time to cook. Add vegetables that take the longest to cook first - dense vegetables such as broccoli and carrots, for example. Next, add softer vegetables such as peppers and onions. Add garlic last.

b) Microwave tips

Cover vegetables to retain moisture that help them cook more quickly, but don't seal tightly. Stir vegetables, especially when reheating, to cook them evenly.

i)   Beetroot: Scrub beetroot gently, but don't peel before cooking. Skins will slip off easily after they're cooked - just peel them while they're still warm. Use paper toweling to avoid staining your hands. Pierce beetroot and other root vegetables with a fork or a skewer to check whether they're cooked.

ii)   Cabbage: Don't overcook cabbage, which would result in a strong, sulphurous odour and mushy texture. Red cabbage would turn bluish-purple while cooked. The cabbage is red to begin with because it contains a high amount of acid. The acid cooks off along with the steam, leaving alkaline cabbage – which produces bluish-purple colour. Hard water, which is more alkaline, would cause more discolouration.

iii)   Cucumbers: Waxed cucumbers should be peeled. Unwaxed cucumbers may be peeled with a swivel peeler or left unpeeled.

iv)   Garlic: Garlic burns easily – add it only at the end of cooking and watch it closely.

v)   Leafy greens: Especially spinach should be washed carefully to remove all dirt. Fill a large pot with water and fresh greens: swish the greens in the water to loosen dirt, which then sinks to the bottom. Wash spinach but don't dry it and cook it with no added water. Greens cook quickly. The water on the leaves is enough if they're cooked over medium heat until leaves just wilt. Less water means more retained nutrients.

vi) Potato: If potatoes aren't organic, it's best to peel them. Peeling potatoes is easiest with a swivel peeler; cut- off any green spots before the cooking. Pierce potatoes (and other root vegetables) to let steam escape – otherwise they could explode in the oven or microwave. When making mashed potatoes, be careful not to "overwork" them. Don't overcook potatoes or mash them in a food processor. This will result in a gluey, rather than fluffy end product. Keep mashed potatoes warm for up to half an hour by putting them in a covered heat-proof bowl. Put the bowl in a pot of hot water on the stovetop with the burner set on low. To keep more nutrients in "boiled" potatoes, boil them for 15 minutes, then drain off the water and steam them until they're done.

vii) Turnips: Turnips should be peeled. Don't overcook them, which will give them an undesirable flavour and texture.

8.9 Deterioration Factors in Vegetables and Their Control

A summary of overall deterioration reactions in fruits and vegetables is as follows. It is important to control their deterioration to increase shelf- life.

a) Enzymic changes

Enzymes which are endogenous to plant tissues can have undesirable or desirable consequences. Examples involving endogenous enzymes include: a) post-harvest senescence and spoilage of fruit and vegetables; b) oxidation of phenolic substances in plant tissues by phenolase (leading to browning); c) sugar - starch conversion in plant tissues by amylases; d) post-harvest demethylation of pectic substances in plant tissues (leading to softening of plant tissues during ripening, and firming of plant tissues during processing).

The major factors useful in controlling enzyme activity are: temperature, water activity, pH, chemicals, which can inhibit enzyme action, alteration of substrates, alteration of products and pre-processing control.

b) Chemical changes

Sensory quality: The two major chemical changes, which occur during the processing and storage of foods and lead to a deterioration in sensory quality are lipid oxidation and non-enzymatic browning. Chemical reactions are also responsible for changes in colour and flavour of foods during processing and storage. Lipid oxidation rate and course of reaction is influenced by light, local oxygen concentration, high temperature, presence of catalysts (generally transition metals such as iron and copper) and water activity. Control of these factors can significantly reduce extent of lipid oxidation in foods. Non-enzymic browning is one of the major causes of deterioration occurring during storage of dried and concentrated foods.

C) Colour changes

i) Chlorophylls: Almost any type of food processing or storage causes some deterioration of chlorophyll pigments. Phenophytinization (with consequent formation of a dull olive brown phenophytin) is the major change; this reaction is accelerated by heat and is acid catalyzed. Other reactions are also possible. For example, dehydrated products such as green peas and beans packed in clear glass containers undergo photo-oxidation and loss of desirable colour.

ii) Anthocyanins: The rate of anthocyanin destruction is *p*H dependent, being greater at higher *p*H. Of interest from a packaging point of view is the ability of some anthocyanins to form complexes with metals such as Al, Fe, Cu and Sn.

These complexes generally result in a change in colour of the pigment and are therefore undesirable. Since metal packaging materials such as cans could be sources of these metals, they are usually coated with special organic linings to avoid undesirable reactions.

iii) Carotenoids: The mechanism of oxidation in processed foods is complex and depends on many factors. The pigments may auto-oxidize by reaction with atmospheric oxygen at the rate dependent on light, heat and presence of pro- and antioxidants.

iv) Flavour changes: In fruit and vegetables, enzymically generated compounds derived from long-chain fatty acids play an extremely important role in formation of characteristic flavours. In addition, these types of reactions can lead to significant off-flavours. Enzyme-induced oxidative breakdown of unsaturated fatty acids occurs extensively in plant tissues and this yields characteristic aromas associated with ripening fruits and disrupted tissues. The permeability of packaging materials is of importance in retaining desirable volatile components within packages, or in permitting undesirable components to permeate through package from ambient atmosphere.

8.10 Nutritional Quality

The four major factors which affect nutrient degradation and can be controlled to varying extents by packaging are light, oxygen concentration, temperature and water activity. However, because of the diverse nature of various nutrients as well as chemical heterogeneity within each class of compounds and complex interactions of the above variables, generalizations about nutrient degradation in foods would inevitably be broad ones.

8.10.1 Vitamins

Ascorbic acid is the most sensitive vitamin in food, its stability varies markedly as a function of environment such as pH and concentration of trace metal ions and oxygen. The nature of the packaging material can significantly affect stability of ascorbic acid in foods. The effectiveness of the material as a barrier to moisture and oxygen as well as the chemical nature of the surface food is exposed are the important factors. For example, problem of ascorbic acid instability in aseptically packaged juices have been encountered because of oxygen permeability of the package and the oxygen dependence of the ascorbic acid degradation reaction.Also, because of the preferential oxidation of metallic tin, citrus juices packaged in cans with a tin contact surface exhibit greater stability of ascorbic acid than those in enameled cans or glass containers. The aerobic and anaerobic degradation reactions of ascorbic acid in reduced-moisture foods have been shown to be highly sensitive to water activity; the reaction rate increases in an exponential fashion over water activity range of 0.1-0.8.

8.10.2 Physical Changes

One major undesirable physical change in food powders is absorption of moisture as a consequence of an inadequate barrier in the package; this results in caking. It can occur either as a result of a poor selection of packaging material or failure of the package integrity during storage. In general, moisture absorption is associated with increased cohesiveness. Anti-caking agents are very fine powders of an inert chemical substance, which are added to powders with much larger particle size to inhibit caking and improve flowability. Studies on onion powders showed that at ambient temperature, caking did not occur at water content of less than 0.4. At higher activities, however, (aw > 0.45) the observed time to caking is inversely proportional to water activity, and at those levels, anti-caking agents are completely ineffective. It appears that while they reduce inter-particle attraction and interfere with the continuity of liquid bridges, they are unable to cover moisture absorption sites.

8.10.3 Biological Changes

a) Microbiological

Micro-organisms can make both desirable and undesirable changes in the quality of the food, depending on whether or not they are introduced as an essential part of the food preservation process or have arisen unintentionally and subsequently have grown to produce food spoilage. The two major groups of micro-organisms found in foods are bacteria and fungi, the latter consisting of yeasts and moulds. Bacteria are generally fastest growing, so in conditions favourable to both, bacteria would usually outgrow fungi.

Foods are frequently classified on the basis of their stability: as non-perishable, semi-perishable and perishable. For example, hermetically sealed and heat processed (e.g. canned) foods are generally regarded as non-perishable. However, they may become perishable under certain circumstances when recontamination can be owing to following. If can seams are faulty, or if there is excessive corrosion resulting in internal gas formation and eventual bursting of the can. Spoilage may also take place when canned food is stored at unusually high temperatures: thermophile spore-forming bacteria may multiply, causing undesirable changes such as flat sour spoilage.

Low moisture content foods such as dried fruits and vegetables are classified as semi-perishable. Frozen foods, though basically perishable, may be classified as semi-perishable provided that they are properly stored at freezer temperature.

The majority of foods (most fresh fruits and vegetables) are classified as perishable unless they have been processed in some way. Often, the only form of processing which such foods receive is to be packaged and kept under controlled temperature conditions. The species of micro-organisms which cause spoilage to particular foods are influenced by two factors: a) nature of the foods and b) their surroundings. These factors are referred to as intrinsic and extrinsic parameters. The intrinsic parameters are an inherent part of the food: pH, nutrient content, antimicrobial constituents and biological structures. The extrinsic parameters of foods are those properties of the storage environment, which affect both foods and their microorganisms. The growth rate of micro-organisms responsible for spoilage primarily depends on these extrinsic parameters: temperature, relative humidity and gas composition of the surrounding atmosphere. The protection of packaged food from contamination or attack by micro-organisms depends on the mechanical integrity of the package (e.g. the absence of breaks and seal imperfections), and on the resistance of the package to penetration by micro-organisms.

Metal cans which are retorted after filling can leak during cooling, permit any microorganisms which may be present in the cooling water, even when the double seam is of a high quality. This fact is widely known in canning industry and is the reason for mandatory chlorination of cannery cooling water. Extensive studies on a variety of plastic films and metal foils have shown that micro-organisms (including moulds, yeasts and bacteria) cannot penetrate these materials in the absence of pinholes.In practice, however, thin sheets of packaging materials such as aluminum and plastic do have pinholes. There are several safeguards against passage of micro-organisms through pinholes in films: because of surface tension effects, micro-organisms cannot pass through very small pinholes unless they are suspended in solutions containing wetting agents and pressure outside the package is greater than within; materials of packaging are generally used in thicknesses such that pinholes are very infrequent and small; for applications in which package integrity is essential (such as sterilization of food in pouches), adequate test methods are available to assure freedom from bacterial recontamination.

b) Macro biological

Insect- pests: Warm humid environments promote insect growth, although most insects would not breed if temperature exceeds 35°C or falls below 10°C. Also many insects cannot reproduce satisfactorily unless moisture content of the food is greater than 11%. The main categories of foods subject to pest attack are cereal- grains and products derived from cereal- grains, other seeds used as food (especially legumes), dairy products such as cheese and milk powders, dried fruits, dried and smoked meats and nuts. The presence of insects and insect excreta in packaged foods may render products unsalable, causing considerable economic losses as well as reduction in nutritional quality, production of off-flavours and acceleration of decay processes due to creation of higher temperature and moisture. Early stages of infestation are often difficult to detect; however, infestation can generally be spotted not only by the presence of insects themselves but also by the products of their activities such as webbing, clumped-together food particles and holes in packaging materials.

Unless plastic films are laminated with foil or paper, insects can penetrate most of them quite easily; the rate of penetration usually directly related to film thickness. In general, thicker films are more resistant than thinner films, and oriented films tend to be more effective than cast films. The looseness of the film has also been reported to be an important factor; loose films being more easily penetrated than tightly fitted films.

Generally, the penetration varies depending on the basic resin from which the film is made, on the combination of materials, on the package structure, and of the species and stage of insects involved. The relative resistance to insect penetration of some flexible packaging materials is as follows.

- Excellent resistance: polycarbonate; poly-ethylene-terephthalate;
- Good resistance: cellulose acetate; polyamide; polyethylene (0.254 mm); polypropylene (biaxially oriented); poly-vinyl-chloride (unplasticised);
- Fair resistance: acrylonitrile; poly-tetra-fluoro-ethylene; polyethylene (0.123 mm);
- Poor resistance: regenerated cellulose; corrugated paper board; kraft paper; polyethylene (0.0254 - 0.100 mm); paper/foil/polyethylene laminate pouch; poly-vinylchloride (plasticised).

Some simple methods for fetching insect resistant packaging materials are : select a film and a film thickness that are inherently resistant to insect penetration ;use shrink film over-wraps to provide an additional barriers and seal carton flaps completely.

Rodents: Rats and mice carry disease-causing organisms on their feet and/or in their intestinal tracts and are known to harbor salmonella of serotypes frequently associated with food-borne infections in humans. In addition to public health consequences of rodent populations in close proximity to humans, these animals also compete intensively with humans for food. Rats and mice gnaw to reach sources of food and drink to keep their teeth short. Their incisor teeth are so strong to gnaw through lead pipes and unhardened concrete as well as sacks, wood and flexible packaging materials. Proper sanitation in food processing and storage areas is the most effective weapon in fight against rodents; since all packaging materials apart from metal and glass containers can be attacked by rats and mice.

8.11 Methods of Reducing Deterioration

Knowledge of deterioration factors and the way they act, including rates of deterioration to a specific category of food, leads to list the ways of lowering or stopping action and saving fruits and vegetables. In order to maintain their nutritional value and organoleptic properties and because of technical-economic considerations, not all the identified means against deterioration actually have practical applications for fruit and vegetable preservation. Technical methods of reducing food deterioration are summarized as follows.

a) Physical

Heating, cooling and lowering of water content drying /dehydration. Concentration, sterilizing filtration, irradiation **and** other physical means (high pressure, vacuum, inert gases)

b) Chemical

Salting, smoking, sugar addition, artificial acidification, ethyl alcohol addition, antiseptic substance action

c) Biochemical

Lactic fermentation (natural acidification), alcoholic fermentation.

This classification of methods of reducing deterioration presents some difficulties because their preservation effects are physical, physico-chemical, chemical and biochemical complex phenomena which rarely act in isolation. Normally, all takes place together or one after the other. These preservation procedures have two following main characteristics as far as being applied to all food products is concerned.

- Some of them are applied only to one or some categories of foods; others can be used across the board and thus are of a wider application (cold storage, freezing, drying/dehydration, sterilization, etc.);

- Some guarantee food preservation on their own while others require combination with others, either as principal or as auxiliary processes to assure preservation (for example smoking has to be preceded by salting).

8.11.1 Preservation by Drying/Dehydration

a) Freezing

Fresh vegetables are dehydrated up to the point where their weight is reduced by 50% and then they are preserved by freezing. This procedure (freeze-drying) combines advantages of drying (reduction of volume and weight) with those of freezing (maintaining vitamins and to a large extent organoleptic properties). A significant advantage of this process is short drying time in so far as it is not necessary to go beyond inflexion point of the drying curve. The finished products after defreezing and rehydration/reconstitution are of a better quality compared with products obtained by dehydration alone. Cold storage of dried/dehydrated vegetables to maintain vitamin C; temperature can vary with storage time and can be at -8° C for a storage time of more than one year with a relative humidity of 70-75 %. Packaging under vacuum or in inert gases is to avoid action of atmospheric oxygen,mainly for products containing beta-carotene. Chemical preservation: a process used intensively for prunes and which has commercial application is to rehydrate dried product up to 35 % using a bath containing hot 2 % potassium sorbate solution. Another possible application of this combined procedure is initial dehydration up to 35% moisture, followed by immersion in the same bath as explained above; this has advantage of reducing drying time and producing minimum qualitative degradation. Both applications suppress dehydrated products reconstitution (rehydration) step before consumption. Presence of desiccants(such as calcium oxide, anhydrous calcium chloride, etc.), especially for powdered products, reduces water -vapour content in package.Preservation by concentration, carried out by evaporation, is combined with cold storage during warm season for tomato paste (when water content cannot be reduced under limit needed to inhibit moulds and yeasts(e.g. aw = 0.70...0.75). Chemical preservation is combined with acidification of food medium (lowering pH); using combined chemical preservatives.

b) Processing

Processing (canning, drying, freezing, and preparation of juices, jams and jellies) increases shelf- life of fruits and vegetables and the nutrients in the vegetables remain available for a long term. Processing includes preparation of raw material (cleaning, trimming, and peeling), followed by cooking, canning, or freezing.

c) Canning

Canning means preserving vegetables by sealing in airtight cans or jars. To home gardener, however, only glass- jars are practical. In home canning, food is heated or processed for a specified time in a closed jar and hermetically sealed with a two-piece cap. Heating jar expels air and halts decay. As the jar cools, lid seals onto the rim and creates a vacuum .All fresh foods contain enzymes as well as naturally vegetablesoccurring micro-organisms including moulds, yeasts and bacteria. The purpose of canning vegetables is to limit these organisms and enzymes.

Ways to can vegetables: Vegetables for canning fall into two categories: high acid or low acid. High-acid foods such as tomatoes are processed using a boiling-water canning method at a temperature of 212°F (100°C). With this method, packed jars are placed in a rack and lowered into a large pot of boiling water. Boiling-water processing is easiest for home canner. Low-acid foods must be preserved with a steam-pressure canner at a temperature of 240°F (116°C). Low-acid foods include green beans, carrots, beetroot, peas and sweet- corn. All canned foods must be processed for specific amount of time required in the recipe to ensure safe product.

Safety issues for preserving vegetables by canning: Steam pressure canning is the only safe way to avoid risk of food poisoning when processing low-acid vegetables. One concern when preserving vegetables is botulism. Home canning recommendations have changed over years: acidification of tomatoes is now advised before canning.

Canning equipment for vegetables: Necessary items for boiling-water canning include: Glass canning jars, Two-piece metal caps (lids and bands), Large canning pot with lid , Jar rack that fits in the pot, Jar lifter (a specialized pair of tongs),Kitchen timer ,nice to have but are not required: Wire basket ,canning funnel

General procedure for canning tomatoes: Wash jars and two-piece caps, dry bands. Heat jars and lids in simmering water (180°F or 83°C) in a sauce-pan - do not boil. Fill canning pot half-full with water, put in canning rack (elevated), cover pot, and heat to simmering (180°F or 83°C) . Wash tomatoes, blanch for 30 to 60 seconds (until skin cracks), dip in cold water .Remove skins, cores, and any green areas; boil tomatoes for 5 minutes. Remove jar from canning pot with jar lifter. Add specified amount of lemon juice or citric acid to jar .Pack hot tomatoes into jars leaving 1/2 an inch (12 mm) of headspace, fill jar with cooking liquid, remove any air bubbles. Dry top of jar, remove lid from hot- water with tongs and place on jar, screw down band. Set jar into elevated rack in canner .When canning rack is full of jars, lower into simmering (180°F or 83°C) water. Keep water level 1-2 inches (25-50 mm) above the top of the jar. Put lid on the pot and bring water to a boil. Process jars for recommended time. Remove jars from canning pot and cool; let them stand for 12 to 24 hours .Check seal and remove band. Store jars in a cool, dry, dark place

d) Pickling and Sauce Making

Another method of preservation where vegetables are prepared with salt and spices; after which they are stored in cans or bottles or jars. Varieties of pickles can be prepared and stored for future use. The shelf- life of pickles is more, and seasonal vegetables can be pickled to make them available in off-season too. Sauces can be made with vegetables like tomato, mint and mustard and stored for a long time. They are made with spices and salt, and sugar is also added to get the sweet- sour taste.

8.12 Decorations from Vegetables: "Fashion in Food"

Food is not only a must for a man, it is also for a pleasant occasion to gather friends and sit around the table. For this event, it is very important to cook delicious dishes and decorate them so to create mouthwatering and exciting appetite of people. The taste and pleasure derived from the meal depends a lot on the visual perception. To make a beautiful composition is not hard if you have a fancy rich imagination. Fruits and vegetables can be artfully cut to be appealing. No matter how wonderful your menu is or if the items are trendy or traditional, individuals "eat with their eyes" before making the choice, and a simple garnish—on the plate, on the pan, or on the buffet table—is a positive enhancement that can show you "really care." And it can also increase the bottom-line. Creative food displays entice appetite, indulging other senses as well. What we see visually appeals to and stimulates the palate. Art of carving vegetables is called 'veggie art'. There are many tools and knives which can help you carve vegetables easily and sharply but as a start there is no need for special carving knives; one sharp-pointed knife is enough to carve vegetables nicely. The knife, however, must be sharp at all times. The list of vegetables which you can think of starting carving include carrot, pumpkin, tomato, cucumber, Chinese radish, eggplant, spring onion, chilly, radish etc.

8.12.1 What is Veggie Art?

Veggie Art is the phrase used when referring to beautiful fruit and vegetable carved centerpieces.

Few tips to carve vegetables: Following points help manage carving of vegetables.

- After carving, fruits and vegetables should be placed in ice -cold water so the petals of flower designs are firm and spread beautifully.
- Carved fruits and vegetables should not be left in water as this may cause petals to become discoloured and spoilt.
- Each type of carved fruit/vegetable should be kept separately. This would save the work in the event that one type is spoilt.

- Store carved fruits and vegetables in containers and place them in a refrigerator, or if no refrigerator is available, by covering them with a damp piece of thin white cloth and putting them in a place protected so they do not dry and wilt.

- After carving, pumpkin should be dipped in water and removed right away. If left in water, flower petal designs would bruise.

Always choose vegetables that are firm and free of significant blemishes. If you cannot serve your veggie art creations immediately, seal them in an airtight container and refrigerate them.

In Asian cuisine firm rules exist for cutting vegetables. On the one hand, it is designed for decorations, something attractive to offer; on the other, it is peeled, split and can be cut so that it is then cooked perfectly blanched, roasted or fried, and made it easier to digest. For a soup, vegetable is shaped differently than a vegetable. For short cooking times in the work, for example, the largest possible diagonal cut surfaces. It cooks faster and still has a little bite to digest better. For soups, cook usually longer. Herbs and spring onions would soon give its active ingredients and thus be chopped as small as possible.

For decoration, cauliflower and broccoli are good, but are not long lasting. Medium- sized cucumbers are particularly straight and are green, have firm flesh and are easy to use. Carrots are available in many shades all -year round. They can be stored in ice -water. Thus, the finished designs remain fresh for longer and crisp. Black olives and olive are suitable for pictures as the eyes of humans and animals. Radishes are very versatile. Also they can be kept stored in ice- water for very long time. From onions, water- lilies can be created. Maize is cooked in small and large pistons, raw or used for decoration.

8.12.2 Decoration Ideas with Vegetables

i)   **Tomato Peeling:** Make a round cut near the scion. Then make a cruciform cut on the bottom. Put a tomato in boiling water for 15 seconds. After that sink a tomato in ice-cold water and peel the skin.

ii)   **Tomato Halves for Stuffing**: Cut a peeled tomato into halves. Remove flesh with a teaspoon. To make a half stable, crop the bottom. Place halves on cucumber slices.

iii)   **Tomato "Petals":** Cut unpeeled tomato into 6-8 equal parts. Remove flesh from the lobules. Place "petals" as a flower or stuff every petal with greens, salads or chopped vegetables.

iv)   **Tomato and Egg Rounds:** Slice a tomato and a hard-boiled egg into 6 slices, put an egg slice on a tomato slice and decorate with pepper.

v)   **Flyagaric:** Crop the bottom of an unpeeled tomato to make it stable and cut a cover from the opposite side. Remove the flesh with a teaspoon and stuff inside with vegetable salad. Put the cover aslant and make white spots with mayonnaise through a confectionery bag.

vi)   **Basket:** Cut -out a basket from an unpeeled tomato. Take out the flesh with a teaspoon and stuff with salads or chopped vegetables.

vii)   **Paprika for Decorations:** Wash paprika and cut a thick piece with the scion. Remove flesh with a knife. Cut into large strips and cut out different things (stars, flowers, leaves) with special molds. These details are great to any composition.

viii)   **Paprika for Stuffing:** Wash paprika and cut a thick piece with scion. Remove the flesh with a knife. To make a paprika stable, crop the bottom. If there is a hole in the bottom, it is possible to cover it with a cucumber or celeriac slice. Stuff it with any salad or vegetables.

ix)   **Stars and Flowers:** Slice boiled carrot or celery finely and cut -out stars, flowers and hearts with special molds. Decorate as one desires.

x)   **Triangles:** Boil peeled carrots and cut long-wise into halves. Then inclining a knife to the left and to the right at a time, small triangles are cut, and can decorate to the dish with them.

xi)   **Sliced Onion:** Different decorations from fresh onion: slice it finely and separate rings carefully; paint the rings while rolling them in ground red pepper or chopped greens.

xii)   **Radish Flower:** Slice a small red radish finely and lay out a flower from rounds. The core can be made from a carrot slice. Make a stalk and leaves from cucumber peel or scallions.

xiii)   **Decorative garnish** is usually made from carrot, turnip and potatoes. To create a beautiful composition, you need big, even and solid vegetables.

xiv)   **Carrot Cones:** Such cones will go to various dishes from meat to vegetable dishes. One can collect a bunch from 3to 5 cones and dress it with parsley leaves.

xv)   Make a cylinder from boiled carrot and round -off the bottom. With a sharp knife make a bow-shaped cut and remove a fine layer of carrot flesh behind it. Cut out scales in staggered rows up. Cone top should have three scales.

xvi)   **Cinderella's Boots:** Such potato boots would be an effective addition to holiday dish. They can be filled with mashed vegetables or dense sauces. Cut out from a large boiled potato a rectangular bar. Keeping the knife in a horizontal way, make a semicircular cut. Then cut out sole and heel of the boot. Round off the edges and sprinkle with breadcrumbs, fried in oil or in a deep fryer. Make a place for a "foot" with a scoop and fill the boot with mashed carrot or green beans.

xvii) Potato Mushrooms: Cut mushrooms would be a fine decoration. Having arranged mushrooms in parsley leaves, would change dish in a better way. Make a cylinder from a boiled potato and round off the bottom. Then cut a potato into two parts and make a hole with a scoop. Collect a mushroom and fry in oil. Sprinkle the cap with saffron.

xviii) Melon Turtle: Take a quite round melon. One half of the melon part is used; add the head and four legs, which are made with pieces of carrot. Design of the turtle's shell is made on to the melon rind. As regarding the tools, you'll need a paring knife and a carving tool to cut the design onto the surface of the melon, or otherwise a lemon zester or citrus stripper can be used, which are usually used to make lines on lemon peel. The first thing is to cut melon more or less in half, one of the pieces is going to be the turtle shell, we're going to cut some concentric circles, like a kind of spider web.

xix) Radish Mouse: Pick a nice radish with oval shape. Trim off green top, leaving a small amount of stem (nose). Cut off a thin slice of the radish to let the mouse sit on the surface (with the tail up). Cut slots for the ears on each side. The slot should be slightly thinner than the ear itself. For the ears you can split the part you cut off earlier into two parts, or better use two slices of another (slightly smaller) radish. Slip the ears into the slots. For eyes cloves can be used or just peel away a small part of the skin with a tip of a sharp knife.

8.12.3 Tips on Selected Fruit and Vegetables for Carving

Onions and shallots should be fresh and without wrinkles. Choose either medium sized or small bulbs of uniform size; Carrots should be straight and of medium or large-size; Radishes should be fresh, firm, and round. Use uniform medium-sized radishes; Chinese radishes should be straight and of medium-size with clear bright skin. The flesh of large Chinese radishes tends to be mealy; Cucumbers should be green, straight, and of medium-size. If the smaller type is used, those with green skin firmer flesh are better for carving than with greenish white skins; Tomatoes should be of uniform size; Plum tomatoes, with elongated fruits, are firmer than round varieties. Choose fresh ones with no wrinkles; Pumpkins should have thick, firm flesh. Such pumpkins have a rough exterior; Chillies should be fresh with firm skins. Generally small ones are used, because if large ones were cut and spread out to make a blossom, they would cover the entire plate. However, large spur chillies are used for making anthurium flowers; Spring onions and leeks should be fresh and green with no yellowing on the leaves. Select thick, medium-sized plants; Cabbage and Chinese leaves should be used fresh with firm, heavy medium sized heads; Lemons should be very fresh; Taro should be of medium size, and should have fine-textured flesh; Cantaloupe melons should be not yet fully ripe. The skin should be pale yellow without wrinkles or scratches; Yam bean tubers used for carving should not be too large. Large, mature tubers have lot of fibres and become frayed in carving;

Watermelons should have red flesh and green rind with no bruises or wrinkles.

8.12.4 Ideas to Make a Vegetable Flower Arrangement

a) Uses of Containers

Use glass containers to display your vegetables. Wood bowls, baskets and ceramic- ware make equally good arrangements, but glass would let the guests admire vegetables from tip to tail. Celebrate spring with a cylindrical container tall enough to hold early baby carrots. Let their feathery tops serve as additional foliage for a handful of white daisies. Fill a rose-bowl with early peas and their pods, and top them with blooming sweet peas.

Radishes can fill the same rose-bowl all by themselves. A single head of lettuce, washed roots and all, makes a spectacular centerpiece--and can still make a great salad the following day.

Use small vases to show off broccoli sprouts in early summer. Their bluish-green color makes a great foil for flowers in all shades of yellow, and a cluster or line of small containers on your table lets you use smallish side shoots from your plants. Cut clusters of beans along with leaves and vine to arrange in a taller vase with snapdragons; allow enough vine to support beans trailing down the side of the vase.

Explore larger containers for summer harvest vegetables. Their thin skins do not welcome water, so keep them dry while having water for flowers. Fill a bowl or platter with tomatoes and summer squashes, leaving room to tuck in a small vase of zinnias, salvia or other summer blooms. Put a plate under a vase filled with herbs, and create a surrounding ring of cherry tomatoes. Accompany a low bowl of summer squashes with a few of their blossoms, placed in shot glasses full of water.

Kales and cabbages offer splendidly sculpted shapes for arranging. Let a handsome cabbage, outer leaves and all, provide a large centerpiece; trim the stalk and set in a plate or low bowl of water. Let the varied bluish, green and purple tones of kale leaves serve as a foil. Add small branches of early-turning fall leaves for even more varied colour. Potatoes, onions, beets and turnips all have fall colour to contribute to arrangements. A tall glass container filled with washed and trimmed root vegetables marks the end of the growing season. Cut toward the stem end of baby turnips and beets. Use toothpicks or bamboo skewers to give them new stems, and add them to flowers arranged in floral foam.

b) Make Vegetable Holiday Trees

Things required are: a Styrofoam cone, toothpicks, green and red veggies (cucumber, cherry tomatoes), cheese, a knife and cutting board.

First all vegetables are cut into circles and squares. Wrap the cone in clear plastic food wrap, securing it with floral pins/scotch tape in one or two places. Push toothpicks into the cone (need to cut toothpicks down to fit in the top of the cone). If toothpicks are angled upwards, this would help keeping veggie slices from falling off. Cover whole cone!

c) Fresh Vegetables can be Good Décor

one needs an assortment of containers, like footed glass bowls, compotes, hurricanes and apothecary jars. Then, pick awesome produce. Look for a mix of shapes - from long stalks, like celery and carrots, to round and plump, like tomatoes and eggplants.

Style I: Pick a tall, thin apothecary jar to hold a bunch of baby carrots with greens. First add small, ripe tomatoes to serve as a base for the arrangement. Next, mix- in jalapeno peppers to add contrast in colour and shape. Finish off by tucking in carrots, letting greens flow down the side of the jar.

Style 2: Select a shallow, footed bowl to be more like a pedestal. First add a bunch of spring lettuce. Next, to it a head of broccoli. The kale lends darker, rich colour and texture. Add petite leaves of radish greens, and finish off by tucking in green onions - bulb and root side out; as they look like curious little flowers.

Style 3: Start with a stalk of fennel to give arrangement some height and movement. Next, tuck in a few yellow squashes for a pop of colour. A portly eggplant is the perfect finish for this grouping. Home garden potential is great; even the smallest patch of soil can produce vegetables, a beautiful display, or both. Assess the site potential, then learn how container gardening, vertical gardening, and specific plants can be deployed in small spaces. Create a plan specific to the garden space. Through home gardening, women have developed proficiency related to plants and garden practices, which help them become better home and environment managers. Their labour is indispensable to maintain the garden and to help keep production cost low. As home managers, women have useful knowledge of numerous domestic needs. By their involvement in the production process, they are able to meet family needs more easily and economically as well as can involve all family- members in this healthy gardening mission and enjoy the tasty harvest of vegetables.

The Authors

Dr. M.L. Chadha, has served as Director, Africa and South Asia of AVRDC-The World Vegetable Centre during 1997 to 2011. He has been coordinating and networking AVRDC vegetable research and home garden programmers for over 20 years, in Africa and South Asia. Before joining AVRDC, Dr. Chadha had worked for 17 years in India in the improvement of vegetable crops, seed production and home gardening. A fellow of NAAS, the Founder Fellow, and Vice President of the Horticulture Society of India. Presently, he is Chairman, Madan Chadha Safe Healthy Vegetable Foundation.

Dr. K.L. Chadha (*Padma Shri Awardee*) is currently President of the Horticultural Society of India and also Adjunct Professor (Hort.) at IARI. He received M.Sc. (Hort.) degree from Punjab University, Chandigarh and Ph.D. from IARI, New Delhi. He had served in various capacities at IARI, New Delhi, PAU, Ludhiana and Central Mango Research Station, Lucknow. He was the Director, IIHR, Bengaluru, Horticulture Commissioner and Executive Director, National Horticulture Board, Govt. of India; DDG (Hort.), ICAR and ICAR-National Professor (Hort.) at IARI.

Dr. Roohani Pal, Ph. D. in Botany and is a gold medalist during her Bachelor's and Master's from Panjab University, Chandigarh. She had been working as a Visiting Scientist with AVRDC-The World Vegetable Center, Regional Center for South Asia (AVRDC-RCSA) based at ICRISAT Campus, Patancheru, Hyderabad, A.P., India, where she has been involved in developing location specific homestead intensive vegetable production modules to ensure nutrition link, their sustainability through small scale seed production and storage. She had also been involved in planning and implementing of adaptive trials on vegetables to identify and popularize superior lines for promotion.

Dr. Satish Kumar Sain, Ph.D. Senior Scientist (Plant Pathology), ICAR-CICR has been working as Assistant Director-Plant Health Management (Horticulture) at NIPHM, Visiting Scientist & Project Coordinator at The WVC- RCSA, ICRISAT campus, Pathologist at Vibha Agrotech Ltd. and Research Fellow at IARI. He has over 17 year experience in R&D and capacity building in the area of plant protection, home gardening, integrated pest management, biological control, ecological engineering etc. in vegetable and other crops.

www.ingramcontent.com/pod-product-compliance
Lightning Source LLC
Chambersburg PA
CBHW050516190326
41458CB00005B/1552